From John's adventure:

John and Sam sat in silence for a few moments before Sam spoke. "Humans are the only creatures on the planet that wonder about why things are the way they are. Man has evolved in such a way as to be uniquely capable of contemplative thought. Only man has the ability to appreciate a sunset as something other than the end of daylight and a daily cue to either begin resting or start hunting. Only humans might wonder why there is a hole in the universe a billion light years across. Why do you think it's weird that you would do something that is, well…so human?"

SPLITTING CREATION
ERASTUS BUCKROD

Copyright © 2016 by Erastus Buckrod. All rights reserved.

This book or any portion thereof may not be reproduced or used in any manner whatsoever without the express written permission of the publisher except for the use of brief quotations in a scholarly work or book review. For permissions or further information contact Braughler Books LLC at info@braughlerbooks.com.

The views and opinions expressed in this work are those of the author and do not necessarily reflect the views and opinions of Braughler Books LLC.

Printed in the United States of America

First Printing, 2016

ISBN 978-1-945091-04-9

Ordering Information: Special discounts are available on quantity purchases by bookstores, corporations, associations, and others. For details, contact the publisher at sales@braughlerbooks.com or at 937-58-BOOKS.

For questions or comments about this book, please write to info@braughlerbooks.com.

Braughler Books
braughlerbooks.com

Dedication

Splitting Creation is dedicated to the memory of my father. Dad was a true force of nature who only saw the best in all people. A man of deep faith and commitment, Dad was fascinated by the world around him with all its mysteries and its majesty. He was the finest example I know of how to live a life of simple gratitude and a child-like enthusiasm for all things. Never wavering from his own ideologies or beliefs, Dad embraced new ideas and never judged others who may view the world differently than he. A shining example for us all, Dad accepted everyone and all things with interest, grace, humor and love.

Acknowledgements

Writing a novel is more difficult than I thought. Without the encouragement and guidance of many people, *Splitting Creation* would never have been possible. The constant loving support, endless patience and wonderful insights from my wife Beth are the singularly most valuable contributions for which I can never adequately express my gratitude. Nothing good has ever happened in our lives for which Beth's hand has not been a steady guide. To Beth, I can only say from the depth of my heart, thank you for all you are in my life.

A number of family members and dear friends have unselfishly contributed their time and intellect to this project by reading and providing comments on both early and late drafts. Their thoughtful input and creative ideas have greatly impacted *Splitting Creation* and my gratitude cannot be overstated to Tim, Becky, Linda, Peter and Bill for their time and helpful comments.

Contents

Foreword xi

Prologue. 1

Chapter 1   Random Events in Time 5

Chapter 2   I-95 13

Chapter 3   A Spring Drive 15

Chapter 4   Sunset 23

Chapter 5   Mr. Mitchell, Meet Sam 29

Chapter 6   First, Let's Do the Math 37

Chapter 7   Upside Down in an Alien World 47

Chapter 8   ET Never Left Home 55

Chapter 9   John and Sam Move On 65

Chapter 10 Neva 69

Chapter 11 Origins 75

Chapter 12 All Things Original Matter So 81

Chapter 13 Life's Basic Elements 91

Chapter 14 Where Life Lives 103

Chapter 15 How Everything is Built 113

Chapter 16 Back in Time 121

Chapter 17 The Greatest Leap of All 125

Chapter 18 One Tree, Many Branches 133

Chapter 19 One is A Lonely Number 137

Chapter 20 Endless Beauty 143

Chapter 21 John and Neva Say Goodbye 157

Chapter 22 Traveling at the Speed of Light 161

Chapter 23	Food for Thought 165
Chapter 24	Sam, Evan and Cherry Pie. 171
Chapter 25	Where am I? 175
Chapter 26	What's On Your Mind Son? 177
Chapter 27	The Bible Tells Me So? 183
Chapter 28	The Phone Call 193
Chapter 29	He's a Soul Man. 195
Chapter 30	I'm On My Way Home. 205
Chapter 31	Parallel Paths. 209
	Epilogue. 211
	Concluding Thoughts from the Author. 213
	Additional Suggested Reading 223
	About the Author 225

Foreword

On Saturday, March 4, 2009 several weeks after I had begun working on an initial draft of *Splitting Creation* a tragedy struck Pittsburgh Pennsylvania, the city I am proud to call home. The following day, March 5th, would be Palm Sunday. For those readers not familiar with the Christian faith Palm Sunday marks the beginning of Holy Week which, is the week leading up to Easter. Palm Sunday celebrates the story of Christ's triumphant ride into Jerusalem on a donkey and the spectacle of Jerusalem's citizenry as they cheered Jesus and paved the way for His parade placing garments and palm branches in His path.

Sadly for Pittsburgh on the dawn of this particular Palm Sunday the city mourned the loss of three police officers murdered in the line of duty just twenty-four hours earlier. Early Saturday morning three officers had responded to a domestic dispute originating in an average, quiet, middle class Pittsburgh neighborhood where homes are well maintained and lawns are manicured. Residents in that neighborhood are unaccustomed to the crime, drugs and violence that permeates many other Pittsburgh inner city and poorer neighborhoods. The day before Palm Sunday a domestic disturbance had erupted between a mother and her son who were having an argument because the dog had peed on the carpet. When the officers approached the home the son ambushed them from his second floor bedroom window, opening fire with a high-powered automatic assault weapon. Two of the officers died instantly from gun shots to the head while a third officer slowly bled to death in the street but, not before he was able to radio for backup. The shooter was an ex-marine who had been dishonorably discharged from the corps. Following a four hour standoff involving the gunman, SWAT team and police, the son was taken into custody alive. When he surrendered the son was wearing a bullet-proof vest and police recovered a stash of weapons, ammunition and supplies that would have allowed the gunman to lay siege to the neighborhood for days, if not weeks, had he been so inclined.

Palm Sunday evening as I scanned local news channels one of our Pittsburgh stations aired a lengthy piece on the impact this tragic event had on the city and its citizens. Because this event happened on the eve of a major Christian holiday, emphasis was focused on the faithful as Christians throughout Pittsburgh offered special Palm Sunday prayers for the victims and their families marking a sad beginning to the Christian Holy Week. At the end of one of the many special news segments reporting various aspects of the shootings a local Catholic priest was interviewed. The priest served the parish where one of the slain officers and his family regularly attended worship services. Responding to the obvious question of how he would help the families and his flock deal with their terrible loss the priest said,

"...obviously, God wanted them (the officers) with Him for this Holy Week, He needed them for something...".

The news report ended with other parishioners giving insights into their belief that this tragic event must be God's will and part of His plan. Friends and acquaintances of the slain officers offered prayers and sympathy for the victims and their families. Several people commented that they knew the families could take comfort in knowing that the officers were now in paradise with God.

This senseless act of violence was disturbing on many levels. As with most events of this type people are left with profound feelings of distress and confusion as they search for reasons why these types of things happen. Equally confusing to me were the comments made by the priest in response to this senseless loss of life. I'm sure the priest felt his words would bring comfort since, like everyone else, the priest grasped at straws to offer some explanation for the inexplicable. It was clear that many people of faith took comfort in the belief that somehow the priest's musings about God's needs made perfect sense because the tragic deaths could indeed be a part of God's master plan.

The priest's explanation as to why three police officers were slaughtered, however, raised numerous questions in my mind. Should we really believe that the three police officers died because God decided He needed them during Holy Week? If this was in fact true, why and for what purpose could God have possibly needed these three men? If God did need these men, why did God decide that the best way to gather them to his bosom was to have them ambushed and shot to death? Was this a spur of the moment decision on God's part? Had God just awakened Saturday morning during Holy Week and decided that He needed some help? Alternatively, had God been planning this for some time? Did God orchestrate the entire event down to the minutest detail in advance or did He simply initiate an inevitable chain of events by having a dog urinate on the carpet and then let the chips fall as they may? If this was part of God's plan, is the man who ambushed the officers innocent because he really had no choice in the matter? Was the gunman simply carrying out God's will, a helpless human pawn in God's greater plan for mankind?"

These are perhaps ridiculous questions to which no answers can be known. That said, I believe it is impossible to invoke "God's will" in reference to such events and not at least consider some of the ramifications this type of thinking engenders. One thing that seems abundantly clear is that without invoking God's will no one can ever know the true reason why people tragically and suddenly end up in the wrong place at the wrong time and suffer as a result. Literally seconds can make the difference between whether someone is witness to an accident or is involved in the accident themselves. Are those life altering seconds in fact part of a divine master plan the creator has for each individual or, are humans simply an evolving life form whose existence on the third planet from the sun unfolds daily in a series of random events?

The story told in *Splitting Creation* is fiction. The scientific principles, concepts and theories employed in the story are factually and technically correct in so far as they were known given the state of our scientific understanding at the time the book was published. New details and truths concerning the universe, physics, chemistry and biology are published almost daily in the scientific literature. Most of this new information is fundamentally sound and has undergone the rigor of strict scientific scrutiny including the requirement that scientific results must be reproducible to be considered valid. This said, I write with a profound recognition that there are secrets in nature and creation that will likely never yield to human exploration and insight. The mysteries of our universe are real and it is essential for science to recognize that some things will never be known. Nevertheless, as our understanding of the majesty of creation is revealed through scientific exploration, perhaps believers and nonbelievers alike might begin thinking about the creator in a different way, in a new light. This new light is not myopically focused and potentially dangerous like a laser which can burn, cut and divide. Rather, this new light is the glow from a cozy fire that illuminates, enlightens, warms and unites us all.

That which is impenetrable to us really exists. Behind the secrets of nature remains something subtle, intangible and inexplicable. Veneration for this force beyond anything that we can comprehend is my religion.

- Albert Einstein (1879 - 1995 AD)

Prologue

Passing its apex in a nearly cloudless azure northern Virginia sky, the sun began its daily decent toward the western horizon. The beginning of the end of another day on the North American continent had begun. Somewhere on the opposite side of the globe at the very same moment in time the very same sun would begin its climb out of the eastern horizon reaching toward its noon height as a full day of events were about to unfold on a different part of the planet earth. Sunrise and sunset. A perpetual earthly cycle repeated unceasingly, simultaneously and without fanfare every second of every day as the third planet from the sun spins on its axis and orbits its sun. That sun, one of a hundred billion stars in one of a hundred billion galaxies that decorate an unimaginably vast and boundless cosmos.

Traffic was unusually light along I-95 despite caution signs indicating the onset of a construction zone several miles south of Fredericksburg Virginia. Approaching the construction zone, a grey BMW 3-series traveling at nearly seventy miles per hour settled effortlessly into the right lane as its driver stole a quick glance in the rearview mirror confirming that there were no other cars behind within at least a half mile. Ahead, a few trucks, an SUV and several passenger cars had already negotiated the changing traffic pattern that ushered both southbound lanes to the left via the Jersey barrier concrete slalom that began more than a half mile ahead of the BMW's current position.

The road construction crews had apparently left for home early to get a jump on the weekend. Light traffic and beautiful weather combined with the obvious in-activity in road work zones allowed most drivers to maintain their highway speed and not slow down as they approached the section of road under construction and the resulting lane changes. As the BMW sped along in the right lane, a tractor trailer carrying a load of Little Debbie Cakes was the only other south-bound vehicle now in sight. Trucks were required to remain in the left lane throughout the length of construction zones and the semi slowed only minimally as it lumbered into the left lane in advance of the approaching lane change barriers.

Dressed in dark suit trousers, shiny black toe-cap shoes and a crisp white shirt with starched cuffs rolled half way up muscular forearms, the driver of the BMW was relaxed, enjoying the quiet, stressless drive and light traffic. Since he had finished work earlier in the day he was now free for the entire weekend. Before beginning his drive south he had removed his suit jacket, folded it neatly and laid it on the rear seat behind him on top of a black leather briefcase that contained some papers and his laptop computer. The top button of his shirt was undone, the neatly pressed collar casually opened at the neck. A light blue striped tie worn tightly knotted earlier in the day was now loosened at his neck and hung limp below a precisely tied four-in-hand knot. As the BMW sped toward the construction zone the driver mused silently, "Could that entire truck really be full of Little Debbie Cakes? How much could a truck load of cupcakes weigh?" A diamond shaped orange sign signaled the pending lane changes one thousand feet ahead. As the BMW approached the lane change its driver checked the rearview mirror once again and was surprised to see a car approaching from behind. Only moments earlier no car had been close and this one was closing fast.

Suddenly more alert and shifting an alert gaze between the Little Debbie semi, the pending lane changes, and the BMW's rearview mirror, its driver knew instantly that the fast approaching car was an Audi which, was easily identifiable by the distinctive four interlocking circles on its front grill. It was also apparent that the Audi was flying since it hadn't been visible in the rear view mirror only a few moments earlier. The Audi continued to rapidly close the gap between itself, the BMW and Little Debbi.

Traveling southbound at more than one hundred miles per hour the green Audi literally had appeared out of nowhere as it barreled down the left lane of southbound I-95. The BMW, the Audi and Little Debbie rapidly approached the beginning of the Jersey barrier and the construction zone lane changes. Little Debbie in the left lane, the BMW in the right lane, the Audi bearing down rapidly in the left lane. The Audi would have to slow down as there was no room for the Audi to negotiate a lane shift before plowing into the back of the semi. Although the BMW was in the right lane, it was not more than a few car lengths behind the rear of the Little Debbi semi so there was no room for the Audi to cut in front of the BMW before all three vehicles entered the lane change area. Either the Audi or the BMW would have to slow down.

Seconds ticked by and the distance between the three vehicles diminished rapidly. As the Audi continued to barrel toward the lane shift zone a sickening thought materialized within the consciousness of the driver of the BMW. The Audi was indeed going to try and cut in front of the BMW as the two cars approached the back of the Little Debbie truck and the pending lane changes.

"*This idiot has got to slow down!*" The driver of the BMW thought to himself and a warm, wet sensation materialized around the open neck of his neatly starched shirt. Tiny, nearly imperceptible drops of sweat appeared on his upper lip and he could feel a bead of sweat break free from his right armpit and run down his side leaving a moist, salty stream of liquid perspiration on his skin. The Audi continued its approach with blinding speed, seemingly oblivious to the pending danger.

At 70 miles per hour or more than 100 feet per second, a mere fraction of a second can mean the difference between a near miss or catastrophe, life or death. One second more, or one second less, to adjust a seat belt or check the mirrors. One second to enjoy a last sip of coffee. One second to smile and recall the sound of children's laughter before turning the ignition and putting the car into gear. One second to admire a sunset. One second to linger and taste a warm, tender kiss or enjoy a fond embrace. The thing is, one never knows before hand whether one second more, or one second less, could mean the difference. One second, 100 feet per second. One second to remind us of how random, fragile and improbable life is.

Time accelerated as the distance between Little Debbie, the Audi and the BMW closed with blinding speed. Rather than slowing down, in the blink of an eye the green Audi seemed to accelerate and it dove from the left lane into the right lane immediately in front of the BMW. Only inches appeared to separate the Audi's front left bumper from the rear of the Little Debbie semi. In a split second the Audi was in the right lane directly in front of the BMW. The BMW's driver braked violently and simultaneously jerked the steering wheel hard to the right narrowly avoiding a collision with the right rear corner panel of the speeding Audi.

With a sickening thud the BMW's right front tire caught the ramp-like leading edge of the Jersey barrier that signaled the start of the lane shifts for the construction zone. The BMW's chassis groaned as the car climbed the concrete wall like a roller coaster hugging the rails through a thrilling corkscrew spin. In an instant the BMW became airborne at more than sixty miles per hour, gracefully rolling its undercarriage skyward as the car inverted, ricocheting upwards and away from the barrier. An eerie silence descended upon I-95 south bound as the BMW floated freely through space. Time ground agonizingly to extreme slow motion. The driver of the BMW looked up through the sunroof, which was actually now down, and he saw that the cloudless blue sky had turned dark and gloomy as the asphalt of the road and the gray concrete barrier on the left side of slalom course enveloped his field of vision. The driver's suit jacket and briefcase floated weightlessly above the rear seat into endless space as a voice in the driver's head silently screamed one last thought, *"What an asshole, didn't he see there was no room to cut in front of me!"*

With a violent, bone crushing jolt the BMW landed hard on its roof atop the concrete barrier on the left side of the highway. The flat, narrow peak of the Jersey barrier clawed at the cabin's interior as the barrier's blunt concrete blade tore through the car's roof. A hellish mixture of sparks, concrete chips, shredded metal and broken glass spewed a dazzling contrail behind the BMW as it skidded on its roof for more than 150 feet along the 9 inch wide apex of the Jersey barrier. At first, the car's forward momentum kept it precariously balanced on its roof until the BMW's slide began to slow causing the car to list precariously into on-coming traffic in the northbound lanes. A grey missile grinding its way along the top of the concrete barrier. As its deathly glide slowed, the BMW teetered along the edge of the barrier. The BMW was listing hard to the left and threatening to drop into the oncoming northbound traffic when a flat bed tractor trailer hauling steal coils and traveling northbound in the left lane clipped the front of the car. The glancing blow from the northbound truck caused the BMW to lurch violently and catapult off of the Jersey barrier. Landing on its roof as it smashed back into the southbound roadway, the BMW bounced hard twice then slid across both southbound lanes, slamming into the concrete barrier that demarcated the right side of the southbound roadway. Like a turtle on its back the BMW continued a decelerating slide. Finally, rocking slightly from side to side while lazily completing a final half pirouette the BMW came to rest on its roof in the middle of the roadway nearly 600 feet from where the green Audi had narrowly negotiated the lane shift. The Audi's driver never saw the accident as the BMW had disappeared behind the Little Debbie truck an instant after the Audi dove into the right lane, perfectly timing the abrupt lane change with split second precision. By the time the BMW had come to a complete stop, the green Audi was nearly a mile further south. The dazzling spring sun continued its lazy decent toward the western horizon with complete disregard to the events that had just occurred on a northern Virginia highway.

CHAPTER 1

Random Events in Time

March 2008. It had seemed that Robert Boorman had been a mechanic for as long as he could remember. After graduating from East Cumberland High School in 1989, Bob entered technical school and became a certified automobile mechanic. Following his graduation Bob was immediately hired by a GM dealership in Baltimore. For a number of years Bob moved from dealership to dealership in the Baltimore area and received several "Service Mechanic of the Year" awards during his ten years with GM. Because Bob was an excellent employee and a skilled technician, he frequently changed service departments at the request of GM to help train and ultimately supervise other new mechanics.

Bob knew cars inside and out. He knew the performance characteristics of every automobile on the road, what they were capable of and their limitations. During Bob's tenure with GM products his eye was always on the opportunity to "move up" to a German car company such as Mercedes, BMW or Audi since he considered these cars to be among the best of the best. When Bob learned through the grapevine that a new Audi dealership was to be opening in Southeast Baltimore, Bob applied for the position of Assistant Service Manager.

At the time Bob submitted his application the new dealership was still months from its scheduled opening. On the basis on his fine reputation, however, Bob got the service manager's job and he was sent to Germany for three weeks at Audi's expense to attend the service manager's training program. The company used its training program to keep its service managers up to date on the latest Audi innovations and to ensure that service at all its dealerships around the world maintained the highest standards. Bob attended the training program with a number of new service managers as well as some seasoned veterans who were there for updates. The trainees had access to all the latest diagnostic tools and equipment in the immaculate state-of-the-art service training facility in Ingolstadt Germany, the site of Audi's largest production facility.

Bob's three weeks in Ingolstadt were some of the best times in his life. Having never been out of the US before, Bob found that he loved everything about Germany. Its beer, its food, its women, and especially its cars. During Bob's stay with Audi in Germany he was housed at company expense in a modest, but comfortable hotel about five kilometers from the training center. The best part of the entire experience, however, was that Bob was given a new Audi A4 to drive during his time in Germany. With the weekends off, Bob took full advantage of the Audi A4 and traveled to Bohn Germany several times on the Autobahn.

The Autobahn is a sleek network of concrete roadways running throughout Germany, Switzerland and Austria. In Germany alone, there are more than 7000 miles (12,200 kilometers) of Autobahn highway, which ranks it as the third longest highway system in the world after the United States and China. The Autobahn's length is impressive given the fact that Germany is barely the size of the state of Montana. American highways simply can't compare. While there are no posted speed limits on the German Autobahn and traveling at speeds in excess of 120 miles per hour is commonplace, traffic fatalities are infrequent compared to other highway systems around the globe. In fact, there are fewer traffic fatalities per mile driven on the Autobahn than all other roads combined in Germany or the US.

There are a number of important reasons behind this interesting statistic of lower traffic fatalities per mile on the Autobahn. For one thing, it's far more difficult to get a driver's license in Germany than in the US. In order to obtain a driver's license in Germany you must be at least eighteen years old and the fees run about $1500.00. In addition, 25 - 45 hours of professional driving instruction are required before an individual can be licensed to drive in the country. In general, drivers in Germany are simply more skilled than are drivers in the US.

Germany and other European countries also take DUI violations very seriously. A driver caught driving under the influence in Germany (blood alcohol level of 0.03 percent as compared to 0.08 percent in the much of the US) will lose their license for two years. At the end of two years, regaining a driver's license in Germany after a DUI is an exhausting and expensive process. Regaining one's license after a DUI in Germany requires prolonged monitoring of blood alcohol levels over months following arrest along with psychological counseling. Most onerous, however, is that regaining one's driver's license costs as much as 15,000 Euros (in excess of $20,000.00) in court costs, doctor visits and attorney fees. In Europe, DUI is taken very, very seriously.

Another thing that makes the Autobahn safer than other highway systems is that it is designed for high speeds whereas most American highways simply are not. Grades are never more than 4 percent and the steepest Autobahn grades all have separate climbing lanes for faster moving traffic. The Autobahn is constructed of freeze resistant concrete and the roadway provides generous shoulders with very long acceleration and deceleration lanes at on and off ramps. Curves are gentle and adequately banked. Reflector posts are spaced at intervals of fifty meters along the entire length of the Autobahn providing excellent visibility even at night. Wildlife protective fencing and crossing areas underneath the roadbed reduce the probability of collisions with large animals such as deer and bear. The Autobahn is also equipped with automated traffic and weather monitoring systems that provide drivers with adequate alerts of accidents and potential travel problems. The Autobahn is a marvel of modern and eco-friendly road engineering that is impeccably maintained throughout its entire network.

Finally, most cars driven on the Autobahn at high speeds are performance cars such as Audi, BMW, Porsche, Volvo, and Mercedes, built to handle speeds that are typical on the Autobahn. They are heavy duty, well built and designed for smooth, quite, reliable performance at high speeds. A German car with one hundred thousand miles is just getting broken in.

During his three weeks in Germany, Bob legally got to feed his passion for driving fast in a nearly new Audi A4. He loved it. When he returned to the US and his new job at the Audi dealership Bob told everyone about what a kick it was to take a great car and drive at those speeds. As weeks settled into months, then years on the new job, Bob thought less and less about Germany and just focused on his role as the new assistant service manager. His boss, Greg Fedan had been a service engineer with Audi for over fifteen years. Bob hoped to step into Greg's job when Greg retired. As a result, Bob went out of his way to make sure his customers were well cared for. Bob worked hard to apply the lessons he had learned at the service training program in Germany to make sure that his service team was the best in the area. It didn't take long for Bob and his team to hit their stride and only four months after the dealership opened and Bob began as assistant service manager, the new dealership was winning acclaims from its customers and Audi for the excellence of its service department. Both of these translated into record sales despite a sluggish economy.

While Bob enjoyed his new role at the Audi dealership, he quickly discovered that the assistant service manager could become involved in all types of dealer activities besides just supervising the service side of the business. Because Bob was an excellent employee held in high regard by the rest of the organization, Bob was often consulted on things like promotional ideas. Bob worked long hours logging plenty of overtime to make sure that the service business remained top of the line. He'd frequently stay late to personally oversee rush jobs that usually involved turning around a customer's car quickly after parts arrived should a car need repairs and the necessary part required was not in the general inventory. Bob never minded the long hours as he loved his job, loved the cars, and loved happy customers. For Bob, the extra time and pay was usually a bonus.

Quite frequently, when a new customer wants a car of a particular color, or with a specific set of options and one dealership doesn't have it in stock, the dealership can log into the dealer computer network and locate a car at a different location. Generally when this happens one dealer will simply trade one car for another and transfer the cash differential on the invoice, if any, through a wire transfer. Usually when this is done, particularly within a convenient geographical area, such as the Baltimore, DC, Richmond or Philadelphia areas, a car carrier tractor trailer will simply pick up the car when dropping off a shipment of new cars and then transport the desired car along with others to another dealer. Occasionally though, when either a customer is in a hurry or there is no carrier making the rounds, an employee from one dealership might drive a car to another dealer, swap cars, run the paper work and bring the new car back to the selling dealership. While this means a customer may get a new car with as much as one hundred miles or so on it, most don't mind and are glad to have their new car sooner, rather than later.

Bob was rarely asked to make one of these swaps since the Baltimore dealership had several semi-retired part-time employees working as drivers who liked nothing more than to take a new car for spin and bring back another new car. Nevertheless occasionally Bob would be asked to handle a pick up, usually at the last minute when other drivers were unavailable.

The internet has changed car buying for both the customer and the dealer. The car shopper can now walk into a dealership loaded with information from the internet already having decided exactly what car, options and price they want. Typically, these sales can happen fast since there may often be little room for price haggling when a well prepared consumer is ready to buy. This is actually fine with a salesman since customers like these are usually shopping seriously. More often than not, they prepared buyer may not even need a test drive or any of the other niceties dealers use to make a sale since they've done all that type of homework earlier. As a result a salesman could theoretically make three or four sales in less time and with less back-and-forth than it takes to get one potential tire-kicker into the car of their dreams. Informed customers can be demanding, however. This was the case with Phil Ventanna who came into the Baltimore Audi dealership one beautiful day in March 2008 with his guns loaded and his checkbook ready. Phil looked forward to buying a new Audi every three years. When he entered the showroom, Phil was delighted to have the opportunity to make the unfortunate salesman, unlucky enough to approach Phil with the standard welcome greeting, wish he had picked some line of work other than Audi sales.

Phil had a general distain for car dealers and particularly their sales staff. He hated the haggling and the mysterious boss behind the curtain. When an offer was made for a car, the salesman would say, "Well, I'm not sure I can do this, but let me go and talk to my sales manager. If I can tell him you are ready to buy the car at these numbers, I'll fight for you." Really Phil thought, who did these guys think they were fooling with this act and the picture of their wife and three kids on the desk facing the customer! Though he would say, if asked, that he hated the thought of playing these sales games and found them a foolish waste of time, if pressed, Phil would have to admit he did get some perverse pleasure out of going toe-to-toe with a salesman and crushing him. He'd already decided exactly what he wanted and what he would pay. He walked into the Baltimore dealership, his financing in hand, and ready to buy. In truth, Phil could barely contain his excitement and couldn't wait to put his haggling skills to the test.

After telling the salesman what he wanted and what he was willing to pay, Phil was disappointed to find that they had already sold the Canyon Red A6 with leather, sunroof, and sports trim package he'd seen on his tour of the lot three days earlier. That car had been sold, however, the dealership had a black one which the salesman told Phil he could drive home that afternoon and that he could make him "quite a deal". Phil wasn't going to hear any of this and said that if they couldn't find him the car he was looking for he'd go elsewhere since he was a busy guy and didn't have time to waste with a dealership that couldn't meet his needs. Since the salesman recognized immediately the well informed customer, he got on-line to see if he could find the exact car Phil was looking for.

Bingo. In less than a minute the salesman had located a car at a dealership in Glen Allen Virginia just north of Richmond that was nearly identical to what Phil wanted. The only difference was this particular car had an upgraded stereo system.

Phil set his jaw and said to the salesman, "I'll take it, but I want to pick it up tomorrow and I won't pay for the stereo upgrade. That's my offer, take it or I'm walking."

Phil had read The Motley Fool's guide to car buying and his offer of 2 percent over invoice for the car without the upgraded stereo was fine with the salesman who actually could have gone even lower had Phil really done his homework. All car sales personnel had special dealer incentives they wore around their necks like a string of garlic to ward off aggressive buyer tactics in order to make customers like Phil feel good about the deal they were getting.

The salesman knew he had the sale and after taking a check for $500.00 from Phil as earnest money, he took the deal to the sales manager who signed off. Phil waited in the sales cubicle ready to pounce when the salesman came back crestfallen that he'd not been able to get the deal, but instead had a counter offer. Much to Phil's surprise, however, when the salesman returned he thrust out his hand to Phil with a big smile and said, "Congratulations, it took some haggling on my part but I got you the deal you wanted. You've just bought yourself a new Audi A6. I'm sure you're going to love this car."

Phil and the salesman completed the preliminary paper work and arranged for Phil to pick the car up at noon the following day. The dealer was going to drive down to Richmond tonight and pick up the car. Service would prep it in the morning and it would be ready when Phil returned the following afternoon. The dealer was happy. The salesman was happy. Phil was elated because he'd just killed them in negotiations.

Although it was a good day for Phil, Bob's day had not started off well at all. He had worked late the previous night and he and his wife had an argument in the morning before Bob left for work. Bob's long hours were taking a toll on his marriage. Bob's wife Betsy was getting tired of dealing single-handedly with the two kids, soccer games, music lessons, teacher conferences and a host of other family things while Bob worked late. Although Bob fixed Audis all day long his family drove a four year old Chrysler minivan and a three year old Ford Taurus. They had no savings due largely to their unwise purchase four years ago of more house than they could afford. Living on a tight financial high wire act kept both Bob and Betsy on edge and sometimes things just boiled over. This morning had started with yet another argument about kids, jobs, cars and money. Bob had promised he'd be home for dinner and they'd talk. As a result, Bob was not happy when the salesman came to him just after 2:00 in the afternoon and asked if someone from service could drive down to Richmond and pick up the Canyon Red Audi A6 since there were no other drivers available and they needed the car for a tomorrow delivery. Bob knew that "someone from service" really meant him since he couldn't pull any of his mechanics off any current jobs. Since they were not going to trade a car for the Canyon Red A6, Bob would have to take one of the dealer loaners and drive to Richmond along with one of the service grunts who washed customer's cars after they were serviced. The grunt would drive the loaner back to Baltimore while Bob drove the new A6. He grabbed Dustin, one of the two kids who washed cars for the dealership. "Come on Dustin, if you have a date tonight you better let her know you'll be late. We're going to Richmond." Bob grumbled.

Bob called his wife and told her he'd be late for dinner, but with any luck and favorable traffic he'd be home before 9:00 that evening. It was a bad time of the month for Betsy. The kids were both sick with spring colds and she didn't feel well herself. "Fine", Betsy spat into the phone and hung up.

At 2:15 in the afternoon on a beautiful March day, Bob got behind of the wheel of the green Audi A4 with Dustin riding shotgun and they hit the road. He was determined to make it home before 9:00 that evening so he could rub that in Betsy's face, playing the martyr's role of the overworked husband who had moved heaven and earth to get home earlier than he had promised. If Bob was to have that satisfaction he was going to have to fly.

CHAPTER 2

I-95

No sooner had they pulled out of the dealership than Dustin started talking about cars and how he and his friends would cruise looking for fun in his 1982 V8 Chevy Impala he was always working on. "Fun" usually entailed driving fast, drinking and looking to hook up with girls. Bob thought to himself, *"One more reason to get to Richmond as fast as I can so I don't have to listen to Dustin's exploits any longer than necessary."*

Fortunately, Dustin was tired from last nights "fun" and when he discovered that Bob was not in a good mood and had no interest in talking, Dustin yawned several times slouched into the leather seat, closed his eyes and was asleep before the Audi pulled onto I-95 heading south. Traffic was light and Bob was flying. His spirits lifted as he sped down the highway knowing that if he could get to the dealership on the northeast side of Richmond before 5:00, he could be home by 8:30 that evening at the latest. He'd already decided to take the A6 straight home rather than drop it at the dealership. This way, he'd save time since they wouldn't prep it until the morning anyway.

Shortly after entering I-95 South, Bob found himself on the Autobahn once again. At nearly one hundred miles per hour the green Audi ate up distances, effortlessly weaving in and out of the light traffic Bob only occasionally needed to slow down when he came upon cars in both lanes. Several miles south of Fredericksburg Bob saw the first construction ahead sign but, traffic was light and moving freely. Bob maintained his speed and shook his head as he looked at Dustin snoring in the passenger's seat and thought to himself, *"Good thing this jerk is asleep as I'd not want him to get the idea he could handle this car on the way back like I can."*

The Virginia Department of Transportation had been working through much of mild Virginia winter to widen and resurface long stretches of I-95 from an area well north of Richmond and extending more than ten miles south toward the outer Richmond city limits. As with all state DOT's, Virginia typically set up new traffic patterns using concrete slabs called Jersey barriers that locked together and created a virtual slalom for drivers as traffic was diverted onto a temporary roadbed while the work on the new surface was completed. In most cases these temporary highways were just wide enough for two tractor trailers to run side by side even though trucks were restricted to the left lane in construction areas. Speeds were reduced to 50 miles per hour in such areas, however, almost nobody paid attention to the posted limits especially when traffic was as light as today. Just as on the open highway where the posted speed was 65 miles per hour most people believe that this means you can go 75 without getting stopped. In fact, if you didn't travel 75 miles per hour in most 65 mile per hour zones you'd most likely get run over.

Traffic was light and the construction crews appeared to have gone home to begin the weekend early. As a result Bob thought it unlikely he'd need to slow down much, if at all, in the construction zone. Nearly a half mile before the advertised construction zone and pending lane changes Bob saw a single tractor trailer ahead in the left lane and what appeared to be a BMW in the right lane. Bob rapidly closed the distance between himself and the trailer truck in the left lane and he saw an easy opening where he could switch lanes, ducking in front of the BMW into the right lane. With seemingly clear sailing as far as he could see through the construction zone ahead Bob's pulse quickened. Judging the distance, timing and speed perfectly Bob dove into the right lane just in front of the BMW without so much as a single apprehensive thought. Smoothly accelerating with the road clear ahead, a quick glance in his rear view mirror told him the BMW was gone and had probably pulled into the left lane behind the semi carrying Little Debbie Cakes. Bob flew past the truck at more than ninety miles per hour. He knew would be home before 9:00 in the evening.

CHAPTER 3

A Spring Drive

A golden yellow sun was the only object in the spring sky on a beautiful, cloudless late afternoon in Virginia. Although the blazing globe that warms the earth had not yet descended far enough to touch the highest peaks of the Blue Ridge mountains, sunset came quickly this time of year as the blue planet continued to change its axis orientation in a manner that would complete the transformation from winter to spring to summer in the northern hemisphere. The rapidly diminishing sunlight gave the northern Virginia landscape a vivid appearance as trees and grass reflected the colors of rebirth, creating a panorama of shadows and various hues of greens and yellows that made everything seem clean and fresh. One could almost taste spring in the air. Although the day had started early for John Mitchell, he was wide awake with anticipation.

Earlier in a nondescript conference room in Washington DC, John and two other managing partners of Eagle Ventures, a Boston based venture capital firm, had concluded a series of discussions with two of the firm's portfolio companies. Having invested a combined total of nearly four million dollars in the two early stage DC-based startup companies, Eagle Ventures met every other month with the company CEOs and executive teams to review product development progress and go over company finances with a fine toothed comb. Eagle Ventures knew that only good management and strict adhesion to tight time lines and aggressive development plans could help insure a good return on the venture firm's substantial investment. Fortunately for the two management teams and for Eagle Ventures, things were progressing smoothly in both companies. In fact, one of the companies was two months ahead of schedule for product launch which was good news to all interested parties.

Following the morning's review meetings the Eagle Venture partners had a quick lunch brought into the conference room while they recapped results from the morning's meetings and tentatively planned the next week or two. Having concluded business efficiently ahead of schedule the partners agreed it would be a good idea for all to begin their weekends early. Two of the senior partners shared a cab to Washington National Airport for return flights to Boston while John got an earlier than expected start on his drive home to Raleigh Durham, North Carolina, a trip of little more than three hundred miles south west of Washington.

Nearly two hours after leaving Washington John was was delighted to have an opportunity to begin the weekend a few hours early on such a beautiful spring day. Yes, spring was definitely in the air. Once he was able to negotiate the DC traffic and get out of Washington proper, traveling south on I-95 toward Richmond John was amazed that traffic was unusually light considering the time of day on a Friday.

NPR had just concluded a report that astronomers had discovered a hole in the universe that was one billion light-years across. An unimaginably gigantic hole seemingly devoid of anything just sitting somewhere in the midst of an endless universe seemingly filled with mostly empty space. No one knew what the hole was or why it was there although this did not stop some scientists from speculating that perhaps the hole might be a portal to another parallel universe just like our own. The NPR report aired a number of sound bites from several prominent astrophysicists who provided various theories as to what the discovery of such a large hole in the universe might mean. Despite many unanswered questions about the giant hole, astronomers and physicists alike were fascinated with the discovery and the possibilities for further scientific investigation it provided. A full report on this immense, yet strangely empty hole in space, was to be detailed in the May issue of the *Astrophysical Journal*.

John wore tailored dark blue suit trousers with a faint grey pinstripe, shiny black toe-cap shoes and a crisp white shirt with starched cuffs that he had rolled half way up muscular forearms. Before beginning his drive south from Washington, John had removed his matching suit jacket, folded it neatly and laid it on the rear seat behind him. A black leather briefcase containing some papers, several company business plans and his laptop computer rested in the footwell behind the driver's seat. The top button of his shirt was undone and the neatly pressed collar spread casually open at his neck. A small curl of black chest hair peaked out above the open shirt collar. The light blue tie that he wore earlier in the day knotted tightly at his neck was now loosened and hung limp below a precisely tied four-in-hand knot.

SPLITTING CREATION

Driving was relaxing and on such a beautiful day the easy drive was helping to melt away the stress of his week. Since 9/11 any business destination that could be reached in five hours or less by driving was considered an easier and less stressful trip by car rather than commercial air. Actually, when you considered the hassle of getting to the airport, parking, getting through security, wondering whether or not your flight was going to take off on time then getting public transportation or a rental car at your destination after landing, a four or five hour drive could actually be faster than a trip by plane. At six feet two inches tall it was not fun for John to be crammed into an isle seat in coach for an hour or more and air travel was not his idea of a comfortable trip compared to four or five hours in his beamer. John tapped the accelerator of the 3-series BMW and easily jumped to nearly eighty miles per hour, flying past two trucks and a minivan with a "baby on board" placard in the window. Smiling to himself as the car responded, he took in the view and delighted in the thought of getting home almost two hours earlier than he'd expected, that is assuming he didn't run into any construction traffic. Such a beautiful afternoon, perfect temperature, sunroof vented and the road clear. John Mitchell would be home for a late dinner with his wife Lucy.

At thirty-six John was the youngest junior partner in Eagle Ventures, a venture capital firm based in Boston. After graduating from Darden Business School in 1996 with and MBA in corporate marketing and finance, John had gone to work as a market research analyst for a privately held software company with a dozen employees in Research Triangle Park, North Carolina. In small companies, most employees typically wore many hats. In John's case because he was a wizard with spreadsheets, quick on his feet and able to put a uniquely fresh spin on company finances and market projections, he immediately became directly involved in building financial models and developing market projections. Along with the company's CEO and CFO, John participated in presenting these facts and figures to venture capitalists. His company had struggled to raise additional private equity financing during a downturn in the economy. At the time, venture capitalists were reluctant to provide funding to a small company with little revenue. Their first software product, a technical marvel that allowed users to seamlessly link multiple different databases and create a new hybrid relational database from results obtained through the query of seemingly disparate information was nine months from full commercialization. John's company was under tremendous pressure to provide a believable timeline for a version 1.0 release of the software product. Assuring potential investors that a market for their product existed that could generate the type of income the company was projecting had proven to be a challenge. Given the skepticism that existed in the venture community at the time, John's company met time after time with usually polite, but always cool receptions by venture capitalists.

The company's largely unsuccessful meetings with numerous venture capital firms were not, however, unproductive for John personally. Eagle Ventures liked John immediately and, seeing his potential, offered him a position in the firm as a financial analyst responsible for evaluating financial projections of small companies seeking to convince Eagle Ventures to invest in the startup's hopes and dreams. Very quickly John distinguished himself not only by working long hours but, more importantly, he displayed an uncanny ability to ask key questions and cut quickly through the fluff contained in most company's financial presentations. The resulting ammunition that John's assessments provided Eagle enabled them to drive company valuations down thereby securing investments in a small number of potential winners under very favorable terms for Eagle. Once the market for Initial Public Offerings (IPO's) opened up again in the late 1990's several of Eagle's companies went public. As a result, Eagle made a killing with a more than a forty-fold return on several of their more successful companies. Eagle treated John well during this time and in late 2007 John became a junior partner in the firm.

Because Eagle was interested in making additional investments in the Research Triangle area and since most of John's work involved travel anyway, John and his family were able to remain in North Carolina. When John was not traveling he worked from home. John's wife Lucy and their two girls Naomi and Jessica were happy to be in NC since Lucy's parents lived relatively close by in Myrtle Beach, South Carolina. Both Naomi and Jessica were very happy in school and had many friends so the ability to stay in the area had numerous benefits which, were only slightly offset by the amount of travel John accepted with pleasure to keep his family in their home. Besides, the climate in Boston could never compare to North Carolina where spring was actually a season and not simply a few days in the week between the end of winter and the beginning of summer.

It was precisely on days such as today that John thought that life simply could not be better. He had an excellent job doing something he loved. His wife and daughters were, all three, precious gems. John's wife Lucy was an adopted Vietnamese orphan. Lucy's father was a Vietnam war veteran who had taken a job with the US postal service after returning from Vietnam. Petite and elegant with fine features and jet black hair Lucy was a stunning beauty. Complimenting her physical beauty was one of the most wonderful and caring personalities John had ever encountered. Although first attracted to Lucy's stunning good looks it was her inner beauty that John fell deeply in love with. John and Lucy met during their sophomore year at William and Mary and dated on and off until the middle of their senior year when they realized they were in love. John proposed to Lucy a week before graduation and they were married in the fall. Their marriage was solid which was easy since they were each other's best friend. Although it was hard for John to imagine a woman more beautiful than his Lucy, their two daughters were even more striking than their mother. The Caucasian-Vietnamese mixture captured the very best from both parent's blood lines and Naomi and Jessica were darling girls. Both had jet black hair like their mother, but thicker and slightly wavy. Almond shaped light brown eyes, small noses and beautiful complexions provided the canvas for endearing smiles. John would be beating the boys away with a club when the girls became teenagers. John was crazy about his girls and was always excited after a business trip to get home to see them all.

Because today's meetings had finished early John had decided to enjoy some rare down time during the drive south on a beautiful spring day. This meant that John had turned his cellphone off and the car's sunroof was open. Even the flashing orange lights atop signs indicating road construction and changes in traffic patterns five miles ahead didn't bother John today.

Lucy was not expecting John until quite late despite his early start and eagerness to get home. Lucy knew that John's meetings typically ran overtime and that Washington's rush hour traffic could be a nightmare. Besides, even though John was now a partner with Eagle Ventures and by all rights he could afford to slow down a bit, long hours had become an all to familiar habit for John and it was quite normal for John to slip in late at night when everyone was fast asleep. Therefore, John hadn't called Lucy yet to tell her to expect him early since he knew that construction and Richmond traffic could cause unexpected delays and throw his best ETA for home out of the window. The girls always got excited if daddy was due home early. It was much easier for the girls and Lucy if Naomi and Jessica weren't expecting Daddy rather than to become disappointed if he got delayed and they had to go to bed before he got home to read a bed-time story and tuck them in. Although he was sure he'd get home much earlier than Lucy was expecting, John would not call with the good news until he was well past Richmond and had a better idea of a realistic ETA.

John was no exception to the general driving public and years of experience had confirmed for him that if the weather was good he could set the cruise control just a hair below 75 miles per hour in the 65 mile per hour zones and run right past a Highway Patrol gunning for speeders. Besides, even at those speeds there were plenty of other drivers who flew past John as if he was standing still. These extreme speeders were the ones that the Virginia Highway Patrol were after.

John's BMW had plenty of get up and go and 75 miles per hour was just a jog in the park for the car. He loved to exercise the vehicle through an almost instantaneous jump to warp speed when there was a truck to pass or he could see in his rear view mirror that he needed to move into the passing lane in front of cars coming up from the rear in order to avoid getting stuck behind some slowpoke going less than 65 miles per hour. The 3-series hugged the road like it was on rails and the effortless power of the German engine and transmission gave John the feeling that he was in control of the entire road, including the other cars and their drivers. BMW's after all were built for the Autobahn where speed of 120 miles per hour were the norm. Seventy-five or even the occasional ninety miles per hour that John reached during his leisurely drive home were not even close to testing the car's peak performance.

Several miles south of Fredericksburg, Virginia John observed that he was about to enter a construction area. About a mile from the start of the actual construction zone and traveling at nearly seventy five miles per hour John settled the BMW effortlessly into the right lane and stole a quick glance in the rearview mirror, confirming that there were no other cars behind him within at least a half mile. Ahead, only a few trucks, an SUV and several passenger cars negotiated the changing traffic pattern that ushered both southbound lanes to the left via the Jersey barriers that created the concrete slalom ahead of John's current position.

As John sped along in the right lane a lone tractor trailer carrying a load of Little Debbie Cakes lumbered into the left lane and was now the only other south-bound vehicle in sight. A diamond shaped orange sign signaled the pending lane changes one thousand feet ahead. As John approached the lane change he checked his rearview mirror once again and was surprised to see a car approaching rapidly from behind. Only moments earlier no car had been close and this one was closing fast.

Shifting an alert gaze between the Little Debbie semi, the pending lane changes and his rear view mirror, John could tell instantly that the fast approaching car was an Audi, which was easily identifiable by the distinctive 4 interlocking circles on its front grill. It was also apparent that the Audi was flying since it hadn't been in view the rear view mirror only a few moments earlier. Traveling southbound at speeds approaching one hundred miles per hour, the green Audi had literally appeared out of nowhere as it barreled down the left lane of I-95. John, the Audi and Little Debbie rapidly approached the beginning of the Jersey barrier and the construction zone lane changes. Little Debbie in the left lane, John in the right lane, the Audi bearing down rapidly from behind in the left lane.

Seconds ticked by and the distance between the three vehicles diminished rapidly. As the Audi continued to rush toward the lane shift zone a sickening thought materialized within John's consciousness. Somehow John just knew that the Audi had no intention of slowing and was going to try and cut in front him as the two cars approached the back of the Little Debbie truck and the pending lane changes.

"*This idiot has got to slow down.*" John thought to himself as the Audi continued its approach on the left with blinding speed, seemingly oblivious to the pending danger. A drop of sweat formed in his armpit and ran down John's side. His hands instinctively gripped the steering wheel harder and his mouth became suddenly dry as desert sand. The Audi was directly on him as the warm, golden sun continued its slow deliberate decent in the western horizon.

CHAPTER 4

Sunset

John Mitchell glanced to his right and noticed for the first time since he'd left Washington a peculiar glow as the sun began to touch the highest peaks of the Blue Ridge mountains in the western sky along I-95 in Virginia. Eerie and seemingly unnatural hues of Carolina blue mingled with brilliant pure white, hot pink, various shades of grey and deep reds, foreshadowing that tomorrow would be another beautiful day along the Atlantic coast.

"Looks like a painting...", John thought to himself as he marveled at the complex pallet of colors to his right outside the window of his BMW. "... and not a very good one at that."

John had little interest in art and the art world had never captured much of John's imagination. While like most anyone John appreciated certain paintings, he would never go out of his way to go to an art gallery unless it was part of a business function. The home offices of Eagle were filled with artwork that different partners had accumulated over the history of the company. Some of this art was valuable and some was not. While tasteful, the firm's art collection was largely an eclectic mix that reflected the diverse tastes of the partners or their spouses.

In truth, John had never paid much attention to art and had only once visited the National Gallery in Washington despite visiting the city no less than a few hundred times over the past five years. As far as his interest went, in John's mind he preferred realistic paintings compared to abstract paintings. Abstract or modern art was too messy, it had no form, and it required that you think about what the artist had in mind in order to understand the painting. In this regard, on rare occasions when John did direct any attention toward a painting his mind tended to wander more toward thoughts of technique than the piece of art itself. To John, it was more interesting to examine a painting and think about the mechanics of how exactly the painter had captured the particular subject rather than to think about the meaning of the work itself. Since John had no artistic talent himself he found it more interesting to ponder how a painter's mind and hand had worked together reproduce the image of an object rather than to think too deeply about the art per se and its "meaning".

In large part, John's minimal appreciation for the artist's technique was the reason that he particularly ignored modern or impressionistic art. As far as he was concerned most modern art didn't require a lot of skill or technique. In John's mind modern art was just color and brush strokes applied at random. Whenever John saw a piece of art that contained no identifiable objects, he immediately was reminded of a program he'd seen on *Animal Planet* some years ago while watching television with his daughters. The program featured several elephants in New Delhi that could "paint". According to the art dealer who sold the elephant's paintings for as much as $1000.00 each in some boutique shops in New York City, these elephants had a unique and innate talent for art. The elephant art promoter insisted that these special animals showed real creativity and actually loved to paint. Using their trunks to hold large brushes filled with various colors of paint the elephants would make long, lazy punches and swipes at a large white canvas. True, the paintings displayed on the program looked just like some modern art paintings that humans had made. However, perhaps the greatest challenge for the handler of the artistic pachyderms had been to reinforce the canvas and easels so that they could withstand the violent thrashings from the painting elephants.

John believed the only difference between elephant art and modern human art was that he was certain the elephants had no idea what they were doing whereas, the human artists may have thought they knew what they were doing. Modern, interpretive art just didn't constitute art as far as John was concerned. He felt sure that if you showed two paintings to some pompous, nose-in-the-air art critic, one done by a human and one by an elephant, there was not a critic alive that could tell which was which.

In contrast, but in a limited way, John could appreciate a painting that looked like a photograph. Precision, proper perspective and subject clarity is what impressed John. Not unlike a well-designed financial spreadsheet, realistic art showed creative skill, precise order and clear functionality. Skill and passion of the creator was readily embodied in the work of a realistic painter causing a sense of awe as the observer may wonder "how did they do that".

The sunset that was just beginning to develop to his right strangely reminded John of some landscape paintings he'd seen that portrayed brilliantly colored skies that just didn't look to be quite real. John simply could not help stealing repeated glimpses of the setting sun as he traveled south on that March evening. Looking at the surreal skies, unnatural colors and mysterious shadows created by the setting sun to his right it suddenly became clear to John how a painter could paint something real and yet have it look as if the artist had embellished upon reality. Sometimes painters really do paint what they see, either with their eye or their mind's eye.

"Interesting what drives someone to create." John thought to himself. "*One artist might paint a sky that intentionally doesn't look real. Another might simply paint what they see and yet, both paintings may look the same. Either way, sometimes reality and the imagination can be the same.*" John's thoughts came to an abrupt pause.

"That's just crazy. What in the hell am I thinking about?" John muttered under this breath with a snort, squinting his eyes and shaking his head quickly to rid such ridiculous thoughts from his mind.

Assuming no delays John had a good four hours driving time remaining before he got home and the lease on the coffee he had consumed in large quantities earlier that day was about to expire. Fortunately, as John snapped out of his subconscious musing about the pros and cons of modern art a rest area sign loomed ahead on his right. In less than minute John had parked his car, got out, and stretched to ease the mild stiffness that had settled into his lower back during the drive from Washington. John rubbed the back of his neck as he tilted his head from side to side then back to front to loosen up his neck which seemed to always become knotted up when he sat for long periods of time without an opportunity to get up and move around. Leaving his car in the parking lot John descended the short flight of concrete stairs that led from the car parking area to the grassy rest area. John walked briskly along the concrete walkway to a brick building that housed the rest rooms and an area with several vending machines. A large grassy expanse to John's right containing numerous mature trees, a few picnic tables and several benches offered a pleasant pastoral setting to weary travelers who needed a quiet, temporary respite from the miles of grey roadway and painted lines that represented their primary world as they traveled the interstate.

Within a few minutes John exited the rest area men's room feeling relieved and invigorated. Ignoring the vending machines John stepped outside into the cool air and breathed deeply, enjoying the clean spring air as he gazed at his surroundings. For the first time since he had pulled into the rest area he became aware that except for him the rest area appeared to be completely deserted. John could hear no one else in either the men's or women's rest rooms. He'd passed no one near the vending machines. His eyes followed the short walkway to the gentle rise and flight of concrete stairs with its grey metal railing that led to the parking lot on his right. Surprisingly there was not another car or truck to be seen anywhere except for John's BMW. Puzzled, John looked to his left where he could hear and see traffic buzzing along I-95 in both north and south directions.

"*Hmmm, did I have to piss so badly that I didn't notice anyone else here when I first pulled in?*" John thought to himself, scratching at the hairline behind his right ear. "*What are the odds that I'd be the only vehicle pulled off at the rest area?*"

"Must be a billion to one." John said to himself in low voice with a slight chuckle. John thought this was amusing because his wife Lucy and the girls always needed to stop frequently at rest stops on long trips. Bathroom stops were guaranteed to consume at least fifteen minutes since with Lucy plus the two girls bathroom breaks were never quick even under the best of circumstances. More often than not, the ladies' room was always more crowded than the men's room, plus, the girls couldn't just go to the bathroom and return quickly to the car. Naomi and Jessica would always stop to "ooh and ah" over an assortment of dogs doing their business and excitedly stopping to sniff everything in sight while their owners walked them around on leashes in the designated pet rest areas. Yes, an empty rest stop, though not impossible, was highly improbable. John would enjoy telling Lucy and the girls about this statistical anomaly given how he always would jokingly complain when they needed to stop on trips.

The setting sun was beautiful. Solar light fragmented into a thousand individual beams as it filtered through the trees, mixing long shadows with brilliant rays of intense light that bathed the surrounding landscape from the edge of the naturally wooded landscape across the asphalt parking lot and into the rest area grounds. Were it not for the cars and trucks that could be heard humming along on the nearby interstate it would have been easy to believe that you were in a park somewhere camping or hiking without a sign of civilization for miles in any direction. The quiet solitude was not lost on John and he stopped briefly to enjoy the moment, taking another deep breath of the crisp late afternoon spring air as he admired the natural tranquility that surrounded him. It was actually nice to be alone and outside in the fresh air if only for a few minutes. John's job was fast-paced. He seemed always either on the phone, reading and answering emails, or staring at a spreadsheet with his nose in his laptop. John was a master at multitasking and could efficiently talk on the telephone and scan emails while flipping back and forth between Microsoft Outlook and an Excel spreadsheet. The email, voice mail, meetings, reports, and analysis never ceased. Work was endless and activities simply flowed continuously like a river of information cascading down a mountain gorge. There were submerged rocks and whirlpools everywhere and it required constant vigilance to keep from getting sucked under and inextricably lost in the current. Narrowing of the information gorge would occasionally cause a normally hectic schedule to become downright frantic and eighty hour work weeks were not uncommon, leaving less time than John would have liked for Lucy, Naomi and Jessica.

At the same time John would readily admit that he thrived on the fast pace. A typical type A personality, John was high energy and the more hectic things became the more he enjoyed his job. In fact, John seemed to think most clearly and perform his best at times when others might become overwhelmed. John reveled amidst ambiguity and chaos. When things were frantic John's mind would automatically focus most intensely in a search for the order and logic that always existed just below the surface of chaos. Order from chaos, that's what John loved most about his job. The more complex and seemingly random a problem was, the greater was the personal satisfaction that came from identifying a solution. The sudden illumination and clarity of thought that broke through chaos like a beacon of warm light was what John always looked for and frequently found.

Nevertheless, despite his ability to thrive on such a hectic work life John would occasionally catch himself day dreaming and wondering what it would be like to be free of the perpetual rat race and just be able to relax for once and simply enjoy being alive. Standing in the deserted rest area, John found himself once again in that rare but, recurring daydream.

"*Boy...*" John thought staring through the trees at the setting sun as it began to sink below the vaulted horizon created by the wooded hills of central Virginia, "*... this is really a beautiful spot. Whoever or whatever made this world did an amazing job.*"

Then, recalling the report he'd heard earlier on NPR, John's thoughts continued to roam uncharacteristically, "*... and what about that hole in the universe that is one billion light-years across they were just talking about on NPR?*

How huge is that!...where did it come from and, what in the heck is it doing there?"

Strange, John hadn't had given a second thought in ages about creation or why anything existed. Since before entering college, John hadn't had much to do with organized religion. Although he and Lucy were both raised as Methodists and were married in a Methodist church, neither he nor Lucy were practicing protestants. Outside of weddings for various friends over the years, they hadn't been to church since their marriage more than nine years ago. Since there fortunately had not yet been any deaths on either side of their respective families, not even a funeral had compelled John or Lucy to visit a church in nearly a decade, save for their marriage.

Although John never thought about it, if he were asked, John would say he was agnostic. By John's definition, an agnostic was someone who believes there must be a God but that no form of organized religion with its creeds, ceremonial trappings and sacramental rituals was necessary to validate belief in the Almighty. At times, John might even catch himself questioning if there was a God at all considering all the bad things that seemed to happen in the world. It was as if no one or nothing was truly in charge. And yet, at that moment, in that place, alone and watching the sunset somewhere along I-95 in Virginia John found himself having what many might call a mystically divine moment. Something inside John stirred as he looked at nature's beauty. He wondered again to himself about a hole in the universe and concluded to that it must be billions of times larger than the earth.

"Oh, its much bigger than that." Came a voice from behind.

CHAPTER 5
Mr. Mitchell, Meet Sam

Startled, John spun quickly in the direction of the voice. There, behind him perched atop a picnic table was a young boy, alone and looking in John's direction. Dressed in faded, torn blue jeans and a grey Nike sweat shirt, the boy also wore a red baseball hat with a black bill turned backwards on his head. Wispy light blond hair protruded haphazardly from under the boy's cap falling over his forehead and partially covering his ears. The boy sat casually on the picnic table with his feet planted firmly on the seat bench. He sported a pair of well worn black Nike Air Jordan sneakers. The boy's thin forearms draped lazily across his knees, his hands hanging between his legs, completely relaxed. Though he had been staring at the same sunset as John just a moment earlier the boy's gaze shifted casually to John, then bounced back and forth between the sunset and John. John hadn't seen the boy when he walked hurriedly to the rest room nor had he noticed him when he came out, which was curious since John was certain that the rest area had been otherwise completely deserted.

"You scared me half to death." John said to the boy. "How long have you been sitting there? And…", John paused for a moment, squinted his left eye and wrinkled his forehead in a puzzled look, then added, "…its bigger than what?"

"Oh, I've been here for quite a while." The young boy said with a slight grin. Then he continued, "I was wondering how long it would be before you noticed me. But, I saw that you were captivated by the sunset so I guess you didn't notice me at the time. Besides, I was just sitting here admiring the sunset myself."

This was a little creepy. John could swear that he had been alone at the rest stop only moments earlier. In fact, John felt as if he had looked around pretty thoroughly when, only five minutes ago, he'd been struck by the fact that the rest stop was completely deserted when he exited the men's room. Looking over his shoulder and again scanning the parking lot John confirmed the absence of any other cars or trucks in the parking area other than his BMW. So, how did this boy get here? Not only that, but the kid was wearing a baseball hat turned backwards. John immediately was suspicious of anyone wearing a baseball cap with the bill toward their back since he thought that made anyone appear at least thirty points lower on the IQ scale. The boy's appearance belied someone who would ever look at a sunset let alone use a word like "captivated"! Who uses the word "captivated" in normal everyday conversation? Besides, what were the chances that any kid would dare speak to an adult wearing dress pants, a white shirt and tie at a rest stop along a highway without another person in sight, unless of course the kid was dealing drugs.

Then again John conceded that he had been a bit distracted when he pulled into the rest stop. The urgency to use the bathroom and his odd thoughts about art, the sunset and a hole in the universe may have caused him to miss the boy entirely. Although somewhat improbable, it was entirely possible John concluded.

John took a hesitant step closer to the boy and was about to ask again what he was doing there when the boy spoke to John, continuing their fragmented conversation seemingly without a second thought and completely oblivious to John's obvious discomfort,

"The hole in the universe. It's much bigger than billions of earths. Our galaxy, the Milky Way, which by the way is one hundred thousand light years across and contains one hundred billion stars, would be just a speck of dust in that hole. Most people don't have the slightest idea of how big one billion light-years is since they have no frame of reference for what a billion is, let alone even bigger numbers. In fact, once you get past a million, most people just go primeval and classify anything larger than a million as just…" The boy brought up both hands, separated about shoulder width and mimed quotation marks twice while saying, "really big". The boy lowered his hands into his lap once again, draping his arms across his knees and smiled,

"Without any idea at all of how a billion compares to a million or a trillion, the concept of these types of numbers causes people's eyes to glaze over. When you throw in light years and add the dimension of distance to what is already an unimaginably big number, sensory overload takes hold and your thoughts turn back to what you were going to have for lunch. But do you know what the really amazing thing is? That hole, one billion light years across, is nothing more than a tiny speck when it comes to the entire universe. Heck, the most distant galaxies that we can see from earth are more than ten billion light years away. And to think that this all started from a single cosmic speck nearly fourteen billion years ago. Every time I look at a sunset like this, it just reminds..."

"Wait a minute!" John interrupted. Holding his hands up and shaking his head while trying not be completely rude. John had the feeling of being suddenly overwhelmed and he needed to stop the boy from going further. John took a few more tentative steps toward the boy and said, "Who in the world ARE you and why are giving me a lecture in astronomy nonsense out here in the middle of nowhere at a highway rest stop?"

"Oh, sorry. My name is Sam." The boy said calmly and without any offense at having been interrupted. "What's your name?" he asked, again flashing a warm smile that showed off a set of perfectly white teeth.

Puzzled, John stepped yet closer. As he approached, John noticed that the boy looked even younger than his size suggested. He might be only ten or eleven and was definitely too young to be alone by the highway at this time of day, or any other time of day for that matter. Nevertheless, Sam seemed completely calm and in no hurry to go anywhere. At just over six feet tall and two hundred pounds, a square chin, rugged good looks and an athletic build, John could be physically intimidating even without intention. Despite what might be characterized as a menacing look on John's face, his brow furrowed and his blue-grey eyes narrowed and alert, Sam didn't appear nervous or concerned in the least as John now loomed over him.

"I'm John, John Mitchell." John replied hesitantly as the crinkle in his forehead relaxed slightly and the look on his face gradually melted into one of curiosity. The left corner of his mouth curled up and his left eye squinted as John tilted his head slightly to the left. The evolving change in John's demeanor and appearance caused Sam to grin once again.

Sam politely nodded hello and said while holding out his hand to John, "Nice to meet you Mr. Mitchell. Are you OK Mr. Mitchell? You look kind of...I don't know...weird. Like maybe you're lost or confused?"

John was definitely not all right. The strange confluence of events surrounding this entire situation had him suddenly a bit unnerved and uncharacteristically cautious. The deserted rest stop, his mental meanderings about art, creation, and the universe had made John's head feel a bit mushy to say the least. Now Sam's sudden appearance in the middle of nowhere and the boy's rattling on about light years, galaxies and holes in the universe had John feeling more than a bit perplexed and somewhat spooked. Still, Sam had been nothing but pleasant and polite, obviously completely at ease with the situation. John was generally quick at reading people and now that the initial shock was wearing off, Sam just seemed to John like a nice kid. True, Sam seemed unusually obsessed with astronomy but there was a quiet confidence about Sam that made him seem wise beyond his years. He was actually quite an endearing little fellow John thought as he took Sam's hand, giving it a firm shake while he peered down at Sam's youthful face. Then without even thinking, John released Sam's hand and simply took a seat beside Sam on the picnic table, stretched his back and neck once again, then draped his forearms across his knees aping Sam's casual pose. John took a deep breath and gazed again at the setting sun as it began to fall beneath the trees on a hill rising gently to greet the western sky.

"You know what Sam?" John paused and thought for a moment, filling his lungs with a deep breath of the crisp Virginia air, then continued, "I'm fine, just fine." And John let out his breath in a long, slow, almost meditative exhale. As he breathed out, the last shred of tension remaining from his meetings earlier in the day and the strange events of past few minutes simply vanished.

"The sunset really is beautiful. It's just that this has been a hectic day, as they all are." John added absentmindedly. "I'm excited to get home to be with my family tonight and just relax. Then out of nowhere I find myself at a rest stop looking at the sunset with some kid I don't know who is jabbering something about billions-of-lights-years-astrophysics-mumbo-jumbo I don't understand and I'm thinking… Wow, creation is really amazing. This is not like me you know? But, here I am Sam, sitting with you and looking at the sunset. Kind of weird huh?"

John and Sam sat in silence for a few moments before Same spoke. "Why would you think that was weird?" Said Sam turning toward John and smiling. "Humans are the only creatures on the planet that wonder about why things are the way they are. Man has evolved in such a way as to be uniquely capable of contemplative thought. Only man has the ability to appreciate a sunset as something other than the end of daylight and a daily cue to either begin resting or start hunting. Only humans might wonder why there is a hole in the universe a billion light years across. Why do you think its weird that you would do something that is, well… so human?"

"For being just a kid, you have a lot of opinions about a lot things." John said. "You think more like a grown up, and an old one at that. Are you some kind of golden child, or maybe a monk or something?" John added with a bit of a sarcastic laugh, then he continued rhetorically without waiting for Sam to respond. "How'd you get so smart anyway? And come to think of it, where are your parents? Shouldn't' you be home doing homework or playing video games or something? Aren't you a bit young to be out here by yourself? What are you maybe ten, eleven at most?"

"Something like that." Sam chuckled. "I'm older than I look I guess. I live just up over the hill behind those woods not far from here at all." Sam pointed across the parking lot directly into the setting sun. "I was just taking a hike in the woods and when I got down here I decided to sit and watch the sun set. This is one of my favorite places to come at the end of the day."

Shrugging his small shoulders, Sam grinned, then continued speaking in his high, melodic voice, sounding almost as if he belonged in boys choir singing a cappella in a large stone cathedral where sound resonated and echoed around every apse and the high vaulted ceiling. "Ever since I can remember, I've been interested in science. I read everything I can get my hands on about creation, the universe and stuff like that. Our solar system, the rules that hold everything together and make life possible, evolution, the Big Bang… everything that surrounds us just seems to me to be the coolest thing. I can't understand why people don't make it a habit to stop at least once a day, look around and say to themselves… Wow! This is unbelievably amazing."

John could only shake his head in disbelief that he was having this conversation with a kid young enough to be John's son. Not often did John ever stop to think about creation in the way Sam was talking about it. Like many people, John managed just fine racing through life on the planet earth with only a rudimentary, and often wrong, understanding of creation, the universe and life itself. When it comes to the technical details of creation, biology, physics, evolution and much of the rest of science most people fall into one of several camps ranging from benign indifference to outright fear.

Some people just believe generally that the nitty gritty details of science are far to complicated for them to understand. News about science such as the discovery of a hole in the universe, while interesting on the surface, is largely incomprehensible for them and therefore completely irrelevant to most people in their daily lives.

Another large group ascribes any details of the mysteries of creation and life to the exclusive and magnificent purview of a particular God upon which their religious beliefs may be grounded. For those faithful, the unquestioning belief that their God made it so is all they need to know.

At the extreme edge of the spectrum of views on science are the strict fundamentalists who believe that scientific inquiry and discovery represent an amoral and evil endeavor. In this regard, science is thought to be one of the many evils in a perpetual war over the battle for man's eternal soul.

John was solidly in the "interesting, but not really relevant" crowd. Yet for some reason he could not fully understand, here John sat listening to a kid with his hat on backwards for God's sake who seemed to have a knowledge of life and a reverence for creation that went far beyond anything John had ever thought about. John was more than a little curious and mildly intrigued by this young boy he'd just met at a quiet rest stop in Virginia along I-95 south.

Smiling as he returned Sam's dazzling and engaging smile John turned his head and looked once again at the setting sun. In the past twenty years it had seemed like John's life had been on a nonstop treadmill. He had been only an average student until tenth grade when he finally kicked into gear academically. Largely the result of hard work, John finished high school with an above average record and was accepted into a number of good colleges and universities. He chose William and Mary in Williamsburg, Virginia where he majored in business and economics, graduating near the top of his class. On graduation, he immediately entered the Darden School of Business at the University of Virginia where he got his MBA and was ranked as one of Darden's top students in his class. Like many MBA graduates from top-tier schools John was recruited heavily by some of the larger investment banks on Wall Street, but he opted for the more risky, yet potentially more rewarding path of joining a small software company in North Carolina. A downturn in the economy led to the demise of the software company but the connections established during his tenure with the small start up led John to Eagle Partners where he had worked his tail off.

Juggling a career, family and extensive business travel had been a challenge. One business deal blurred ceaselessly into the next like waves rolling onto the shore. While John's life was financially comfortable by all accounts, his mind boarded on frantic most of the time. On the one hand John relished the hectic life. Nevertheless, John realized that he could not remember the last time he just sat and looked at a sunset and he certainly couldn't recall a single instance where he'd given second thought to why things were the way they were. That was until today when John's mind seemed, for whatever reason, free to wander. A peculiar confluence of events had brought him to this picturesque spot. Now, sitting casually on a picnic table at rest stop along I-95 John had met a strange, intelligent and thoughtful boy.

John took a deep breath and started to speak but, stopped.

"What is it Mr. Mitchell?" Sam asked. "Were you going to say something?"

"Sam." John began. "For starters, how about calling me John." John smiled at Sam and gave him a friendly wink. "And second, how about telling me something about the universe and this giant hole that you seem to know so much about?"

"Are you sure Mr. Mitch… I mean John?" Sam said with a somewhat skeptical look. "Don't you have to get going? I can go on and on for a long time once I start talking. Sometimes I can't shut up and people just walk away. Not that I blame them." Sam added.

"I'm sure." John replied, "I've got some time to kill so you go right ahead. My family is not expecting me until late tonight. Since I got a very early start I can still be home much earlier than I had thought when the day started. Besides, a little more time to enjoy the sunset would do me good. I promise not to just walk away, ok? If you begin to get boring, I'll let you know."

John looked at Sam, smiled said, "So Einstein, tell me about the universe."

CHAPTER 6

First, Let's Do the Math

Having been granted John's permission to launch into a discussion of his greatest interests, Sam began talking to John, man to man, displaying a maturity and breadth of knowledge that left John with his jaw in his lap. Though just a boy, Sam spoke with a quiet authority and sense of reverent wonder that engaged John fully and kept his keen attention on everything Sam said. Sam's relaxed demeanor evolved into excitement and enthusiasm as he began to talk passionately about a range of topics. Although his young soprano voice had yet to mature, Sam's voice had a resonance to it that was melodic and hypnotizing. As Sam talked he drew John into him with both the sound and cadence of his speech as well as the words themselves. John found himself hanging on Sam's every phrase, drinking in the wealth of information that sprang from this youth. A blond haired fountain of knowledge with his hat on backwards.

Sam began with the simple topic of numbers which, at first John thought was silly since John was a professional numbers guy. John quickly learned, however, that Sam was talking about numbers in a way that was far beyond John's realm of comprehension.

Since graduate school John had lived and breathed finance and corporate spread sheets. For the past ten years he had been immersed in budgets and sales projections for start up companies. Often financial projections for even young start up companies ran into the hundreds of millions of dollars depending upon how unrealistically optimistic company executives chose to be. The truth was that if a venture capitalist didn't see at least tens or hundreds of millions of dollars in a company's financial growth projections and the company didn't have at least a remotely reasonable plan as to how they would achieve those types of numbers, most venture capitalists would simply not be interested. John knew numbers, how they worked, and how double-digit growth rates for investments could yield huge returns for savvy investors and their partners. What John was not prepared for, however, was Sam's way of describing and using numbers to help John understand for the first time just how fascinating and BIG, REALY BIG, the universe actually is.

Sam explained to John that most people have limits when it comes to absolute numbers. Both extremely small or exceedingly large numbers become blurred once a certain number of zeros is passed. As a result, concepts of size, time and distance become mind-bendingly gigantic once that numerical threshold is reached. Numbers containing lots of zero's are simply difficult to get one's arms around making it nearly impossible to truly comprehend both the vastness of the universe as well as the infinite smallness of sub-atomic particles. Unapologetically, Sam pointed out that even people like John who work with math all the time simply just don't comprehend the types of numbers required to describe the size of the universe or, the "cosmos", which, Sam defined for John as what we see at night when we look up. Similarly, Sam also pointed out that people have difficulty comprehending numbers needed to define the microscopic cellular, molecular and atomic universe that makes up everything we can touch or see. As a result, Sam began by talking to John about numbers in a way that brought size and distance into better focus.

To begin, Sam provided John with some basic facts about the cosmos. According to the best scientific estimates, the universe began with something referred to as the Big Bang nearly fourteen billion years ago and has been continually expanding ever since. The cosmos contains hundreds of billions of galaxies. Each galaxy contains, on average, one hundred billion stars. Our galaxy, the Milky Way, resembles a giant spiral one hundred thousand light years across and sixteen thousand light years thick at its center. Within the Milky Way the closest star to our sun is Alpha Centauri which is a little more than four light years away. In contrast, the next closest galaxy to the Milky Way is called the Andromeda Nebula which is over two million light years away. The furthest galaxies from the Milky Way that we know about are more than ten billion light years in the distance. Amazingly, although the cosmos is comprised of hundreds of billions of galaxies collectively containing trillions of billions of stars, 99 percent of the entire cosmos is seemingly empty space. However, as John would learn later, this empty space is not really as empty as it appears.

"You see, numbers like a hundred billion or a trillion billion are nearly impossible to fully comprehend. Most people simply don't think about numbers in a way that allow them to grasp how much larger one billion is than one million, or one hundred billion is compared to a billion." Continued Sam as

SPLITTING CREATION

he sat looking off into the sunset with John. "Nope, big numbers get lost and it becomes easier to lump things into the 'really, really big' category without any understanding of what that means. To most people the earth seems huge. However, in the grand scheme of things the earth is just the tiniest most infinitesimal speck in a universe that is, for all intents and purposes, endless."

Sam went on to explain that the reason people don't understand numbers like these is not that people are incapable of such understanding. Rather, people don't have a frame of reference that allows them to intellectually comprehend the expanse of time, size and distance necessary to define the cosmos. This is not difficult to understand since everyone typically lives largely within a world defined by their own personal universe. For example, it would take John about 20 seconds to walk from the picnic table where he and Sam were sitting to reach his car in the parking lot of the rest area. In another 4 hours John would arrive home assuming an average speed of 55 miles per hour. If he needed to fly to San Francisco on business, it would take John 5 hours in the fastest commercial airplane traveling at 600 miles per hour to cover the 3000 mile distance from the east coast to the west coast of the United States. If you are in a hurry or haven't traveled much, 3000 miles in a speedy 5 hours can seem like a very long distance and a terribly long time. If most of your trips are 10 miles to work and back, traveling between the east and west coast of the United States is a very long way to go indeed. Likewise, 5 hours is more than one-fifth of a day, so spending 5 hours doing just about anything consumes a meaningful portion the day. In some places, the Grand Canyon is a mile deep and a mile wide. Standing on its rim and looking across to the other side, one cannot help but gasp at seeing how big the Grand Canyon is. Yet the Grand Canyon is quite tiny compared to the planet earth which is 25,000 miles in circumference at its equator and an average of nearly 4000 miles from the earth's surface to the planet's molten core. Our sun is more than one million times larger than the earth and yet, our sun is far from being one of the largest stars in the universe. Therefore, when Sam talked about billions of years, and hundreds of billions of galaxies each containing hundreds of billions of stars, to say nothing about a hole in the universe that is a billion light years across, these concepts of size, time and distance seem to John as nothing more than....well, just really, really big.

Sam helped John grasp the real meaning of numbers like one hundred billion by turning the conversation to something that John new well. Money.

"So…" Asked Sam, "… have you ever thought about the number one billion in terms of dollars John?"

"Sure I have, all the time." John said with cocky grin. "Numbers are my job. I meet nearly every day with young companies that may someday be worth hundreds of millions of dollars. The richest people in the world are worth billions. I know that a billion dollars is one thousand million dollars which is a lot of money."

Sam smiled and said, "Yes John, you clearly know what a billion dollars is in the abstract sense, but have you ever thought about what a billion dollars means in real, practical terms? Think about the giant hole in the universe which we were talking about earlier. It is one billion light years across. How big is that really? Your heart beats approximately 100,000 times each day or nearly 32 million times in just one year. That seems like a lot, but in 80 years, your heart will not beat more than 3 billion times. When you try to think about a galaxy that contains 100 billion stars, what does that mean in terms of the total number of stars in the cosmos since the Milky Way is just one of hundreds of billions of galaxies?"

"I guess I'd never thought of things being that massive." Said John. "That's a lot of money or heart beats isn't it?"

"Yes it is. " Sam answered, then continued. "So to begin with, think about one billion for a minute. Suppose that on the day you were born you were given one billion dollars in cash. A nice inheritance wouldn't you say? However, this one billion dollars comes with two minor stipulations. First, you could not invest the money, but rather, just keep the money as cash under your mattress. The second requirement was that you needed to spend every dime of that one billion dollars by the time you died at the age of eighty."

"Eighty!" John moaned. "I was planning on living to be at least one hundred!"

"Ok, we'll go with a one hundred" Sam smiled. "Clean living, exercise, lots of vitamins combined with excellent genes and you could make it to one hundred I'm sure. So, how much money would you need to spend every day of your life in order to spend one billion dollars in cash over a lifetime of one hundred years?"

John did some quick math in his head and answered proudly, "That's about $30,000.00 a day."

"That's a good estimate although the real number is a bit less," corrected Sam before continuing with a sly smile. "Its actually $27,397.00 a day. To put it another way, that's about 27 cents for every beat of your heart. With that much money to spend every day you would need, for example, to buy a new car, and pay for it in cash every day of your life. That's 10 million dollars a year which is a considerable amount of money for nearly anyone. It's far more than most people will see over their entire lifetime even if they saved every penny they ever earned."

Sam went on. "Now think about 100 billion stars in the Milky Way and convert that into money. If you were given 100 billion dollars the day you were born instead of just one billion, you'd need to spend more than 2.5 million dollars (actually $2,739,700.00) every day of your life, or one billion dollars each year for one hundred years to spend all your money! By the way, that's like shelling out about $27.00 and some change every time your heart beats."

John smiled and thought to himself, *"Bill Gates and Warren Buffet really do have a lot of money."*

Sam went on with yet another analogy that is useful for thinking about numbers like one hundred billion. Seemingly having information effortlessly on the tip of his tongue, Sam casually spewed out, "You know John there are 31,560,000 seconds in one year. The Milky Way galaxy contains 100 billion stars. That means, that if a star were to explode (in other words, die) once every second in the Milky Way the galaxy would lose more than 31 million stars in one year. That's about one exploding star for every single beat of your heart. While 31 million is a massive number, it represents only 0.03 percent of all the stars in the galaxy. Therefore, if one star in the Milky Way died every second for the next one hundred years it would hardly make a difference that anyone would notice…that is unless one of the exploding stars happened to be the sun! Looking at this another way, if a star died once every second in the galaxy it would take more than 3000 years before the last star in the Milky Way blinked out. Of course, this doesn't actually happen since the universe is not static. Rather, the universe is in a state of perpetual creation. Stars are dying and new ones are being formed all the time in all the galaxies that decorate the known universe. Indeed, every second there are literally billions of stars dying and billions of new stars being being created throughout the entire universe in the hundred's of billions of galaxies that make up the cosmos. Death and birth are continuous throughout creation and this balance between life and death of stars occurs in numbers that are nearly impossible to fathom."

"So when you hear a number like 100 billion, which is the number of stars in a galaxy, you need to think about that number in terms that give you a frame of reference as to how many stars that actually is. Then, just think about 100 billion galaxies, each containing on average 100 billion stars and you may begin to get a better idea of the shear vastness of the cosmos. There are more stars in the universe than there are grains of sand on all the beaches of earth." Sam said with a proud grin. Taking a deep breath, Sam puffed his chest just a bit.

John interrupted, "That giant hole in the universe we were talking about is one billion light years across? That's immense? Sam nodded while John continued and asked, "So I have a feeling your going to tell me that a light year is long way all by itself isn't it Sam, let alone one billion light years!"

Sam nodded, "Yes, a light year is a very, very, very long way to travel."

Hardly stopping for a breath, Sam continued excitedly using hand gestures that would make any Italian proud. Sam next explained to John about the speed of light and the great distances that exist in the cosmos. The hole in the universe is unfathomably large by any stretch of the imagination. Even if the hole was only one billion miles across, it would be far too enormous for nearly anyone to grasp. However, this hole is one billion light years across, meaning that a billion miles is a mere speck compared to the size of this giant, seemingly empty hole recently discovered by astronomers.

Light travels at the blinding speed of 186,000 miles per second. For reference, the circumference of the earth is about 25,000 miles around at the equator. This means that in one second (count: one thousand one!) light would travel around the earth more than seven times. This didn't seem that much to John until Sam did the math once again and explained that in one minute, light travels more than 11 million miles. In one day, light travels more than 16 billion miles. If the hole in the universe were a paltry one billion miles across, light could traverse the entire expanse of one billion miles in about 90 minutes. But, a light year, or the distance that light travels in one year is an astonishing 5.8 trillion (5,800,000,000,000) miles! This hole in the universe, representing only the tiniest part of the entire cosmos is one billion light years across which is 5.8 trillion billion (5,800,000,000,000,000,000,000 ,000,000) miles. So, even at the speed of light, it would take one billion years to traverse this empty hole in the universe.

Again, Sam took John on another leap from hundreds of billions to the trillions and reminded John again of the 27 million dollars he would have to spend each and every day for one hundred years to use one hundred billion dollars in his life time. Yet an even greater number to conceptualize, 5.8 trillion miles is the distance that light travels in one year. In dollar terms, John would need to spend $2,150,000,000.00 (2.15 BILLION) every day, or $90,000,000.00 (ninety million dollars) every hour, or $1,500,000.00 (one and a half million dollars) every minute of every day for one hundred years to spend $5.8 trillion dollars in his lifetime. An unimaginable number in terms of size, but yet, it represents only the tiniest fraction, one billionth, of the distance across the giant hole in the universe that Sam and John had been talking about…a dizzying one billion light years across. The nearest star to our sun is more than 4 light years away and our entire galaxy is a mere 100 thousand light years across. Yet these vast distances represent literally a drop in the ocean compared to the massive hole scientists had discovered in the cosmos.

In order to help John better understand distances and numbers that ballooned into the trillions Sam led him through a thought experiment involving travel. The speed of the fastest commercial jet liner is about 600 miles per hour. In order to cover the same distance in a commercial jet that light travels in one year John would have to make 2 billion trips between New York and San Francisco. At the speed of a commercial passenger jet this would take more than one million years of constant, unending travel. Even if were possible to cruise along at the speed of the Apollo moon capsule, which at 40,000 miles per hour is the fastest manned vehicle ever flown, it would still take more than 150,000 years to cover the distance that light travels in just one year.

John was awestruck. He'd known that the speed of light was fast and that the universe was big, but he had no framework of comprehension as to what "big" meant when it comes to the universe. Finally, using some simple references to money and travel, Sam had helped John glimpse just a bit more clearly what numbers like one billion, one hundred billion and one trillion actually meant. With his new perspective, John reflected back on the news report he'd heard about the discovery of a hole in the universe that was one billion light years across, and suddenly, he felt very, very small.

"Sam." John said. "We…I mean our earth…" He paused looking at the vivid color pallet of the sunset as his mind raced to grasp the significance of what he was thinking. "…we are nothing. Earth is just the tiniest speck in the midst of something that is so unfathomably grand that it is nearly impossible to comprehend. It makes one wonder why we are even here and where this all came from."

"Ahhh, you've finally gotten to the essential point John." Sam's said, his voice becoming softer and his mood noticeably more contemplative. "You see, humans have the unique ability to think about their actual existence. At least as far as is known, humans are the only creatures on the planet possessing a sense of consciousness that allows them to expand their world beyond the present and that which can be seen, touched, smelled and felt. Many animals including bees, ants, apes, dolphins and whales, for example, can communicate in sophisticated ways with members of their own species and, some are no doubt highly intelligent creatures. Nevertheless, only human's actually think about things such as why they are here and how immense the universe is. This is what gives human beings the ability to intellectually grasp the size and meaning of creation and know that they are a part of something that is far greater than themselves. It causes, or at least it should cause, a profound sense of awe and wonder inside every person that no other species can enjoy. Creation is an astounding thing John. Why was the cosmos created in the first place? How was it created? And, in the midst of a limitless expanse of space, time and distance, how and why did the earth and life in its limitless forms along with human consciousness come into existence at all?"

John looked at Sam with astonished bewilderment. Sam had just put into words exactly what had been just now running around in his head. The young boy with his hat on backwards sitting beside John on picnic table in the middle of a rest stop in Virginia had read John's mind. John felt as if he was talking with someone much older and wiser than himself.

John leaned back stretching his back then took a deep breath, sighed heavily, shook is head and just smiled as he leaned forward and again rested his arms on his knees. A few minutes passed as he and Sam sat in silence before Sam continued. "Its hard to know if the earth is unique in the cosmos. The physical events and perfect alignment of a countless number of coincidences that resulted in the earth being formed in just the right way, in just the right orbit around the sun, at just the right time, are highly improbable. Some might say so improbable that our earth and everything on it must have been created for some purpose. On the other hand, there are those who would argue that considering the uncountable number of galaxies in the cosmos containing an even greater number of stars, the improbable actually becomes highly probable simply by pure random chance. In reality, given infinite time and infinite possibilities, even the highly improbable is likely to occur once if not many times purely by random chance. If your chances of winning the lottery were only one in one hundred million and you entered one hundred billion lotteries, you'd win the lottery a thousand times purely by random chance. With more than a hundred billion trillion stars in the cosmos, some could argue that there are most assuredly other worlds like earth in existence."

"I wonder what the chances are for that?" John pondered out loud.

"Well, in fact…" Sam answered with a knowing smile, "… I'm glad you asked!" Sam continued, taking obvious pleasure in another chance to talk about something he loved. "There is, in fact, something known as the Drake equation named after a famous American astronomer, Frank Drake. In 1961, Drake attempted to offer a way to calculate the number of planets in the Milky Way on which intelligent life may exist as on earth. The Drake equation predicts that in the Milky Way there are at least one thousand planets with intelligent civilizations. While conceptually logical, the Drake equation might be thought of as quite absurd since it multiplies together seven different factors, each one an uncertain estimate, to come up with the number of potentially intelligent civilizations in the galaxy. Each of these different factors is estimated using conventional wisdom to offer a best guess at parameters such as: a) the number of stars that are born each year in the galaxy, b) the fraction of those stars that have planets, and c) the fraction of those planets that might support life. In reality, however, the number can really be anything since there is no way of verifying all of the estimates for the seven factors in the Drake equation. In spite of the practical uselessness of the Drake equation it is interesting to think about the possibilities for the existence of other worlds where intelligent life may exist. In that regard, despite the uncertainty of the seven factors in the equation, what the Drake equation does do is verify that there must be intelligent life somewhere other than earth because no matter what estimates you use for the seven factors the Drake equation yields a number far larger than one. Most certainly it is extremely arrogant for humans to believe that earth is the only planet in the entire cosmos where intelligent life exists."

"I've often thought the same thing." John said. "Even without having an appreciation for the numbers, it would seem like there must be life on some other planet somewhere in the universe. Besides, a number of seemingly intelligent people claim to have seen UFO's and even to have been taken aboard flying saucers by aliens."

"Oh come now John, you don't seriously believe that UFO's have visited earth do you?" Asked Sam with a somewhat surprised look. "Especially after coming to appreciate the time and distances we've just talked about."

"Well, sure." said John. "For all I know, you could be an alien. After all you seem to know so much about everything. And you do wear your baseball cap backwards."

John and Sam both laughed and Sam patted John on the shoulder saying, "That's a good one John."

John shielded his eyes as he squinted into the sunset. Endless, ever changing hues became more vivid as the gigantic orange ball set the March evening sky ablaze as the sun continued its inevitable lazy descent below the western horizon.

CHAPTER 7
Upside Down in an Alien World

Earl Peters had been driving eighteen wheelers for more than twenty years and, other than hitting a few deer and a raccoon or two, he had never been in an accident involving another vehicle. Earl also had never gotten a ticket for any moving violation in all his years on the road, a truly remarkable record for any driver private or commercial. For the past nine years Earl had been a long-haul driver for McKee Foods Corporation out of Stuart's Draft, Virginia. McKee Foods was the parent company of Little Debbie Cakes and for most of his tenure with McKee, Earl hauled Little Debbie Cakes and other sweet delights up and down the east coast from Maine to Florida.

The day had started like any other for Earl. After his trailer was loaded Earl left Stuart's Draft around noon to avoid morning traffic and was headed south on I-95 with a load of snacks and desserts. Earl would make multiple stops at McKee distribution sites as far south as Miami. He would be gone five days and would cover a distance of nearly two thousand miles on this run.

Earl liked the quiet solitude of the open road and prided himself on his clean driving record. Twice in the past nine years he'd been named "Driver of the Year" by McKee. An honor recognizing Earl for never having missed a day of work, consistently being on-time for deliveries, and his spotless driving record.

Traveling south on I-95 at a fairly steady seventy miles per hour Earl had only moments earlier checked his right rear view mirror and saw a grey BMW pull into the right lane a comfortable several car lengths behind his truck. Traffic had been relatively light so, while he was always attentive, Earl's mind wandered as he took in some of the beautiful Virginia landscape. He was relaxed and making good time.

In an instant, Earl caught sight of the green Audi flying by him on the right. Puzzling to himself *"where in the hell did he come from?"*, Earl checked his right mirror again and was surprised to see that the BMW had vanished.

"Must have pulled right in behind me to let the Audi pass." Earl thought. In twenty years of driving Earl had seen all kinds of drivers on the road and having an Audi pass him at ninety plus miles per hour was nothing terribly unusual. Almost instantaneously a semi heading north bound on the opposite side of the Jersey barrier began laying hard on his horn. This caused Earl to glance into his left mirror just in time to see the north-bound semi jerk violently to its right, striking a glancing blow on what looked to be a car upside down on the Jersey barrier directly behind Earls rig! It all happened so fast that Earl could not be sure what was going on in that brief instant. Looking again into his right mirror, however, he could now see clearly that a car had just landed on the road behind him and careened off the right side of the Jersey barrier. The car skidded on its roof and spun on the pavement as sparks, glass and twisted metal flew everywhere in a sickening shower of destruction.

"Holy shit!" Yelled Earl as he instinctively down shifted and began applying his breaks hard bringing the big rig to a screeching but controlled stop nearly five hundred feet ahead of the mangled car. By the time Earl had brought his rig to a complete stop and turned on his emergency flashers the car was also at a full stop in the middle of the road, on its roof and blocking both south-bound lanes. Grabbing his cell phone Earl leaped down from the cab and began racing toward the car dialing 911 emergency as he ran. While he was waiting for a connection Earl took a few seconds to wave his hands over his head in an effort to alert oncoming drivers of the blocked roadway.

As Earl neared the overturned car, traffic was already slowly beginning to pile up behind the accident and was now stopped for a several hundred yards in both north and south directions along I-95. Cars and trucks were stacking up like beads on a string as each one saw the stopped cars ahead, slowed down, and came to a halt. Drivers strained their necks as they peered out of their windows to see what was going on. Other than Earl, nobody had yet stopped long enough in the south bound lanes to brave getting out of their vehicle. The north bound truck on the other side of the interstate that had clipped the BMW was stopped with its emergency flashers on, blocking the north-bound right lane. The semi driver was standing behind his rig waving his arms in the air as a signal for on-coming cars to slow down. Vehicles heading north could still pass the semi in the left lane, however, there was some debris on the road making it necessary for traffic to slow to a crawl in order to avoid the glass, plastic and metal strewn across the roadway. The inevitable rubbernecking coupled with the debris obstacle course on the road brought northbound traffic to a virtual standstill. Southbound, both lanes were blocked completely as motorists had nowhere to go.

Earl was within one hundred feet of the overturned car when the 911 operator came on the line. Giving his position and telling the operator that an ambulance and tow truck would be needed, Earl described the accident as a bad one and let the operator know that traffic was blocked in both directions. He gave the 911 operator his cell number, hung up and clipped the cell phone back into his belt holster as he slowed to a cautious trot, bending down slightly so that he could see into the mangled passenger compartment of the overturned car. Delicate wisps of white smoke were beginning to rise from the engine compartment of the overturned car. Earl bent down further and continued his approach in a crouching gait as he walked quickly, but cautiously to the driver's side of the car. Finally coming to a halt Earl peered tentatively in to the shattered window of the driver's door. Inside, Earl could see the limp airbags that had deployed as a result of the accident. The body of a man hanging upside down in the driver's seat was clearly visible. The body was held in place by the seatbelt and looked like a rag doll, motionless, limp and lifeless.

Drifting in and out of consciousness and experiencing fleeting moments of believing he was dreaming, the injured driver had no idea where he was or whether it was night or day. There was a strange sensation of pressure on his shoulders and abdomen. His face was warm but his hands and feet were ice cold. Although he felt as if his eyes were open he could see nothing except an aura of color that seemed to be everywhere in his field of vision. Regardless of whether he believed his eyes were open or he tried to close them and clear his head by squinting hard, the images of color would not change. He could hear no sounds other than the ever-present voice of "self" that is in everyone's head with which, he seemed to be having a conversation about nothing in particular. Trying to get his bearings in this silent multi-colored world, he attempted to lift his head in order to have a better look around. This, it turned out was surprisingly difficult. Finally and with great effort, he raised his head but, curiously, his chin seemed to touch his chest. "*Odd.*" He thought, since he was sure he was raising, not lowering, his head.

"Mister, mister."

The driver's attention suddenly snapped away from his inner self as he thought he heard a distant voice.

"Mister, mister."

Unable to identify from which direction the voice was coming it seemed that the voice was calling to him. It was then that he noticed that his arms were raised over his head and he had the sensation that he was reaching skyward.

As Earl peered into the driver's side of the BMW he saw the driver attempt to move his head as if to look around. "Mister, Mister." Earl shouted. "Can you hear me?"

Steadying himself with one hand on the undercarriage of the car, Earl knelt further and peered more closely into the car so he could get a better view of the car's interior and the driver who was held in place by the seatbelt as he hung up side down in the driver's seat. There were no other passengers in the car. Earl could see an open briefcase wedged between one of the rear headrests and the crushed roof of the car. Contents of the briefcase were strewn around the inside of the vehicle. A suit jacket hung by the loop in its neckline from the rear garment hook if it had been casually placed there intentionally, out of the way and neat. Earl could tell from looking at the car that the driver had to be badly injured and was, in fact, surprised to see the injured man attempting to move.

Instinctively, Earl got up and tried to open the driver's door but it was jammed tight and several hard tugs by Earl told him that nothing but the jaws of life would get that door open. The window on the driver's side was completely shattered and as Earl got down on both knees he found he was able to get his head and arms partially into the driver's compartment so he could reach in and touch the driver's shoulder.

"Don't move." Earl ordered the driver. "You've been in an accident and you are hurt. We are going to have to get you out of here. Help is on the way." Earl explained in a voice as calm as he could manage given the circumstances.

By this time several other southbound motorists who had been stopped first by the accident had left their vehicles and were running toward the overturned car. Fortunately at the moment of the accident, thanks to the unusually light traffic there had been no one within a quarter mile of the accident site so other vehicles traveling southbound on I-95 were able to safely stop. Earl had seen his share of accidents and he immediately took charge of the situation. Without removing his hand from the injured man's shoulder, Earl pulled his head out of the driver's compartment and looked up at the gathering group of on lookers.

"He's alive!" Earl shouted.

The injured man was now certain that he heard voices. The voices were telling him not to move and he was sure he heard "help!" Trying once again to raise his head he felt his chin on his chest. Still not sure if his eyes were open or closed, a gentle hand on his shoulder gave him a sense of comfort despite feeling that he was completely alone.

"We are going to need to get this man out of here." Earl shouted perhaps louder than he needed. Although an eerie silence had descended on the usually noisy interstate and he did not need to yell to be heard, adrenaline had kicked in and Earl was feeling the need to take action, and fast.

"This car could catch fire at any minute. I called for help but there is no guessing when it can get here." Earl continued. "Do any of you have any medical training? This man is badly hurt."

Two other men were now kneeling beside Earl and peered into the overturned car's passenger compartment while another man and two women stood hesitantly several yards away. Reflexively, the man and women who were standing took a few steps backward when they heard the word "fire".

Apparently, nobody had any medical training that they were admitting to since no one stepped forward when Earl had asked. As the two men knelt beside Earl, one said, "We'll help you get him out."

The other man nodded and said, "I'll go around to the other side and see if I can reach his seatbelt and help from there."

Speaking over his shoulder to the gathering crowd of onlookers Earl asked if anyone could find some blankets and maybe some water.

Earl squeezed his head and burly shoulders back into the driver's compartment in an effort to assess the situation further while one of the two other men circled to the passenger's side of the car. The passenger's window was also obliterated. As Earl wrestled to push the limp airbag out of the way, he could see that the driver's feet were more or less hanging under the dashboard as if the driver were suspended from a pole using gravity boots, but otherwise, his legs appeared to be free. The exterior of the overturned car where the northbound semi had clipped it was badly mangled. Fortunately the rigid passenger cage designed to protect occupants from rollover and crush injuries had remained amazingly intact. Other than the roof being bowed slightly at its center and shattered glass fragments everywhere, the passenger compartment was in fairly good condition.

"I think I can reach his seatbelt release!" Exclaimed the man who had gone to the passenger side of the car. He was now halfway in the car, lying on his back and looking up at the driver who had now stopped moving and was hanging lifelessly still strapped into the driver's seat.

"OK!" Said Earl. "It's going to take several of us to hold him so he doesn't fall like a stone when you release his belt. We don't want to hurt him any further, but we need to get him out of here as fast as we can." Earl added.

The injured man wakened again from of his dream-like state as a searing pain shot through his back and he thought he heard a distant shriek.

Turning onto his back and now looking directly up at the driver, Earl could not notice that smoke, now a mixture of white and dark grey, was beginning to spew more heavily from the engine compartment. Earl could hear an approaching siren, either an ambulance or police car, although it sounded as if it was still some distance away.

"I'm going to try to support his legs!" Said Earl.

Then talking to the man on his side of the car Earl continued "See if you can hold him by his shoulders while this other guy releases the seat belt. Once the belt is released we need to get it out of the way and lower him down so we can bring him out headfirst through this window. OK, ready? On the count of three."

The man on the side with Earl reached into the driver's compartment a far as possible but, could not get his head inside because Earl's girth filled most of the driver's window. Nevertheless, the second man was able to grab the driver's left shoulder and had his other arm across the driver's chest, grasping the driver's right shoulder as best as he could given the tight quarters.

Pink, red, orange, grey and black again burst into the injured man's field of vision as he tried once again to raise his head only to have his chin touch his chest. "*This just isn't right.*" He thought as he fought the sensation that he was in a dream from which he wanted to waken, but could not. Nothing seemed right in his dream world. The man could hear voices again and he felt as if he was floating up, up, up, down, down, down. One, two, three…

The seatbelt released easily and all three men strained as dead weight now was fully in their hands. The third man on his back on the passenger side of the vehicle helped Earl hold the injured man's right leg and hip and with his other free hand tried to maneuver the belt out of the way. Straining, the second man on Earl's side of the vehicle immediately began to lower the driver's shoulders directing them toward the driver's side window.

"Sorry." The man holding the injured driver's shoulders said as his elbow dug into Earl's groin.

"Just watch his head…" Said Earl, adding, "… don't worry about me."

By now a few other men who had been standing close to the overturned car ran over to assist. Grabbing any part that belonged to the injured driver, pulling, lifting, and dragging, many hands helped to get the injured driver out and as far away as possible from the car. Heavy black smoke now billowed from the engine compartment of the mangled wreck. Someone had found a few blankets and the injured man was placed on several of these that had been layered to form a crude mattress. Just then, a police cruiser came to a stop on the side of the narrow roadbed as near to the accident site as was possible given the line of stopped traffic and the narrow berm in the construction zone. A police officer jumped out of the cruiser, grabbed a fire extinguisher from his trunk and raced toward the overturned smoking car and the growing crowd that had gathered at the accident scene, many of whom now surrounded the injured driver.

"Is there anyone else in that car?" Asked the officer.

"No, he's the only one." Said one of the men who had helped carry the injured driver to the blanket.

"Wait here." Said the officer as he sprinted to the smoking vehicle. Earl and the man on the passenger side were just standing up after extracting themselves from the overturned car when the officer arrived. The man who had helped with the injured driver's shoulders had stayed with the injured driver helping the others transport the limp body to the blankets. "Are we sure there is not any one else in there?" Asked the officer as he knelt down to look in while Earl stood up, gingerly brushing broken glass from his shirt, bare arms and the front of his jeans.

"No, just the driver for certain." Said Earl. "Is he ok?"

"Get back!" Ordered the officer as he aimed the fire extinguisher at the engine compartment and blasted the smoking undercarriage with foam from the red metal canister. After the fire extinguisher was exhausted the officer turned and jogged back to Earl who, along with the other men who had helped, were now crowded around and looking down at the injured man laying on the blankets.

"Move back please. Is anyone here a doctor? The Ambulance should be here in a few minutes." Said the officer to no one in particular. He bent down over the injured driver to assess his injuries and see if there was anything he might do before the ambulance arrived. Turning his attention briefly away from the man laying on the ground the officer spoke over his shoulder at Earl.

"You did a brave thing to get this guy out of the car. It might have caught fire at any moment. I'll need to get some statements from you and anyone else who witnessed the accident."

The officer had enough medical training to know to feel the man's neck for any sign of a pulse. "He's alive!" Said the officer.

Floating over a multicolored landscape, the searing pain in his back gave way to an even worse pain in his abdomen. His eyes cracked open and he tried to focus. What he saw was frightening. He was lying on his back, naked on a cold steel pallet. He shivered. Standing all around him looking down were people…no, not people! They looked like people but they were not. Huge almond shaped black eyes containing no pupil stared out from the center of elongated oval shaped faces with tiny mouths. One of the creatures extended a thin hand toward the man and pressed firmly against his neck. Searing pain shot through his entire body from head to foot as an explosion of color burst into his mind's eye like fire works on the fourth of July. The sudden explosion of brilliant light and vivid color blinded him and the images of creatures evaporated as his body was poked, prodded and violated. He tried to scream for help but, in this dream, no matter how hard he tried he could make no sound. He was trapped in a nightmare from which he could not waken.

CHAPTER 8
ET Never Left Home

Once John innocently brought up the idea of aliens, Sam's unnaturally sky blue eyes began to sparkle as he comfortably stepped through the door John had opened, launching into a far reaching discourse on the concept of aliens and the visitation of earth by beings from solar systems far, far away.

Descriptions of strange objects in the sky have been around nearly since the dawn of man. However, the modern UFO debate was catalyzed by the proclaimed observations of a recreational pilot named Kenneth Arnold who in June 1947 reported seeing what he described as a "flying saucer" while Arnold was flying from Wyoming to his home in Oregon.

On the one hand it may be arrogant to believe that earthly humans might be all alone in a vast and virtually infinite universe. Nevertheless, most modern religions foster the idea that humans have been uniquely created and have a special personal relationship with the creator. More fascinating are the ideas of a creator that not only intervenes in the daily lives of individuals, but that the Almighty promises to reward the good and punish the bad as well. While this might be viewed as a good thing, having no other intelligent or similarly "blessed" life forms in the entire universe would mean that the full attention of the creator was focused exclusively and, presumably intently, on earthly humans. Some might argue that this is a rather myopically self-important point of view considering the tiny speck of cosmic dust upon which we travel through the cosmos. At the same time, this is often the very argument used to emphasize the special relationship that humans have with a particular deity. Believers will marvel that considering the vastness and complexity of the universe the creator takes a specific interest in them personally and that the faithful can share a one-on-one relationship with the Almighty. Believers have no difficulty with the concept that man is entirely alone in the universe nor do they question the idea that humanity was created purposefully by a supreme being who knows them, loves them, and cares for them as individuals. In their minds, every human being is known and loved by God from the very moment of conception when an egg and sperm unite and that this relationship continues throughout all eternity. These believers would argue therefore that it is impossible for earth to have been visited by extraterrestrials from other worlds since man is singularly God's ultimate creation and aliens therefore simply do not exist.

In contrast, speculation on the existence of other worlds and visitation of earth by intelligent beings from other planets is consistent with the belief that we are not alone in creation. A number of scientists, archeologists and philosophers have provided intriguing arguments that the earth may have been visited in the distant past by intelligent beings from other worlds. Numerous books have been written on this subject detailing similarities in the beliefs and artwork of ancient civilizations around the earth. Ancient civilizations such as the Egyptians, Mayans, Incas, to name just a few, crafted images depicting human-like forms that bear a remarkable resemblance to modern day astronauts often sitting in what appears to be some form of space craft cockpit. Similarly, many civilizations and religious texts refer to gods that were too frightful to look upon and who that had the ability to fly through the air seemingly at will. Theorists claim that such alien visitations were responsible for many of the great leaps forward by ancient civilizations including: a) construction of pyramids and other mammoth stone structures around the globe, b) perpetual and amazingly accurate astrological calendars that detail and predict the position of the sun, moon and stars both past, present and future, c) precise mapping of the earth's land masses including the continent of Antarctica before it was covered by ice, d) transoceanic navigation and, e) the transfer of similar beliefs systems and technological capabilities across the globe to geographically widely distributed ancient cultures who could not realistically have had any physical contact with one another without some form of alien intervention.

The alien concept even extends to the debate on the origin of life on earth itself. This, Sam pointed out to John, is not to be confused with the theory of evolution. Charles Darwin's theory of evolution addresses the natural changes and progression of species through a process of natural selection over vast periods of time on earth. In contrast, the origin of life centers on a debate that attempts to explain how life on this planet came into existence in the first place. Prior to the origin of life, there would have been nothing to evolve. Ergo, the origin of life comes first, then Charles Darwin and the theory of evolution follows.

SPLITTING CREATION

Alien theories on the origin of life on earth abound and posit in one form or another that life was intentionally planted on the earth by cosmic travelers. Some argue that aliens actually created man along with other animal species through genetic experimentation. Alternatively, other theories hypothesize that life arrived on earth purely by accident as a single cell or biological molecule spawned on some distant planet that miraculously hitchhiked through the cosmos on a meteor. Eventually, by pure random chance, this hitchhiker life-form crashed into the earth and found a fertile environment where life gained a fragile foothold and later thrived. Ideas such as these, however, create a circular argument on the origin of life by raising the question of how and where did the life form that rode a meteor to earth arise? On and on the argument goes without end.

Ideas of alien life forms populating the earth stand in stark contrast to the purest origin of life theories which argue that life on earth simply began spontaneously, the inevitable result of eons of random chemical reactions that took place in a primordial soup. These random and undirected chemical reactions eventually gave rise to crude biological molecules that over the millennia became living creatures. Random chance theories of this ilk are all based upon the concept that nearly anything is statistically probable within the known laws of physics and chemistry considering the incomprehensible span of time since the Big Bang. In other words, statistically speaking, life had to originate somewhere, so why not on our earth. Given the countless number of stars and possible planets involved even the most remote possibility may be rather commonplace in terms of absolute numbers. Therefore, some would argue it is actually impossible that life would not have spontaneously formed on earth as well as on countless other similar planets throughout the Milky Way and the one hundred billion other galaxies that populate the cosmos. It's simply a numbers game where the quantities are so large that literally anything is possible.

Still other concepts provide even more esoteric, yet fascinating fuel to the debate over the existence other worlds. For example, some of the laws and concepts of modern quantum mechanics are consistent with the idea that there may be parallel universes. One of the most convoluted and mind bending concepts of particle physics known as quantum entanglement postulates that every subatomic particle exists in two but opposite forms at precisely the same time even over infinite distances. While this is very difficult to grasp conceptually, it essentially means that if you look at an apple and see it as red, another identical apple somewhere else in the cosmos is seen as green at the very moment you are seeing the red apple.

Theoretical concepts of particle physics are consistent with the existence of an infinite number of parallel universes which are identical in every way down to the minutest detail. In those infinite parallel universes an infinite number of Sams and Johns are sitting on a park bench at a Virginia rest stop having a word-for-word conversation that is identical in every way. These countless parallel universes may continue into the future in exact step for infinity. However, it is speculated that certain events may cause one universe to split resulting in the instantaneous creation of a completely new parallel universe. The new parallel universe is identical to the original universe in every way up to the event that resulted in the sudden creation of a new universe. In the new universe Sam and John would travel different paths in a parallel galaxy, in another universe, in some other dimension. Parallel universes may be either light years apart and spread across the furthest regions of the cosmos, or they may exist literally only fractions of an inch from one another.

There is no way to prove the existence of parallel universes other than through mathematical calculations and creative experimentation as our understanding of quantum physics evolves. Regardless, some scientists speculate that holes in our universe like the one reported on NPR or perhaps black holes, which are the remnants of dead stars, represent doorways or portals to other universes or other dimensions through which, sophisticated space and time travelers may pass to visit other worlds.

All of these ideas, either alien, parallel universe, or biological roulette fly in the face of the belief of most religions. As a result, the interface between science, pseudoscience and religion has formed the battleground of some of the most intense debates and conflicts in modern society. It was within this very context that Sam and John talked about aliens visiting earth.

After Sam had framed the parameters of the alien concept he said, "So John, think for a moment about aliens from another planet and just what that might mean in the practical sense. You know that the United States is currently considering a program to return to the moon and from there, launch a mission to Mars which is the next nearest planet to earth. Interest in Mars is intense since it is speculated that of all the planets in our solar system, Mars is the one that is most likely to have supported some form of life."

"Yes, I know." Said John, shaking his head and continuing, "Given how large the universe is, Mars doesn't actually seem so far away from earth any longer."

"That's the point." Sam said. "Even though Mars is not far from earth in the cosmic sense it will be a tremendous technical and logistical challenge to get fuel, food, air, water, etc., for a mission to travel the forty-seven million miles from earth to Mars. When man traveled to the moon in the 1960's, the Apollo capsule reached speeds of forty thousand miles per hour which, is the fastest man has ever traveled. Even at forty thousand miles per hour, it will take months to make the one way trip from Earth to Mars. And, once the astronauts landed on Mars they'd have to be able to eventually return to earth so there would need to be mechanisms in place to either carry enough fuel and supplies for the return trip or somehow produce these while on the surface of Mars."

"Now, just suppose that there were a planet with intelligent life on it that is circling the next closest star to our sun, Alpha Centauri, which is four light years from earth….that's nearly 24 trillion miles. At forty thousand miles per hour, it would take almost seventy thousand years to travel from Alpha Centauri to Earth."

"That's impossible." Said John, then added immediately. "But, what about an alien space ship that could travel much faster?"

"Well…" Sam said, "… suppose they did. Even if it were possible to increase the speed of travel by one hundred times to four million miles per hour for example, it would still take seven hundred years to make the journey."

Looking toward the sunset, Sam said, "Even if it were possible to travel at the speed of light, it would take four years and that assumes the pilots would not crash into something along the way! But the more important thing is that this assumes that there is a planet orbiting Alpha Centauri that has intelligent life capable of that type of technology. What if it is not Alpha Centauri but some other solar system somewhere else, not four light years away, but four hundred, or four thousand, or forty thousand light years from earth somewhere in the Milky Way? Or what if this alien is from another galaxy all together several hundred million light years away? When you understand the distances and the reality of the numbers involved it becomes logistically improbable that an alien could make the trip from their planet to earth. Not only that, but according to most alien theories, the aliens have visited earth repeatedly many times over the years. This brings up all kinds of interesting questions. Are these round trips to and from the alien's home planet or do the aliens have a base camp somewhere closer to earth that allows them to make multiple short trips? If they have a base camp where did they get the materials to build and supply it? Do they transport the materials from their home planet or do they harvest resources from earth or some other planet they stopped at along the way? There are not roadside rest stops in space."

"Some may argue that alien life forms have found a way to travel the great distances of the universe by traveling faster than the speed of light or by taking short-cuts through worm holes or even folds in the fabric of space. This still raises all kinds of logistical questions but more importantly, if an alien has the ability to undertake this type of sophisticated travel why would they visit only remote areas of earth and abduct lone individuals rather than land in populated areas and communicate directly with our most sophisticated scientists and political figures? Seems a bit strange to go to such a great effort and travel such a long way just to buzz a few cow pastures, snatch a random country boy, then fly off."

"Furthermore, how would aliens even find us to know we are here?" John piped in with an enlightened look on his face. "Seems to me that would be harder than trying to find a needle in a haystack, it would be like trying to finding a needle in a billion haystacks! It would be….well, next to impossible for an astronomer in another galaxy to identify earth in the first place. I remember now reading something about the fact that we are using our most powerful telescopes to try to identify planets orbiting other stars that might be capable of supporting life. I'm sure this is not an easy task and even if we do find one or more other planets that appear to be earth-like how would we know exactly whether or not they contain only germs or, in fact may harbor some other form of alien life that might be looking back at us through their telescopes!"

Sam smiled a warm grin, looking like a proud parent who has just seen their child take her first steps. Sam's grinning encouragement emboldened John to continue, "I can just see it now. Mr. President, by pure dumb luck, we have stumbled on a small speck of dust orbiting a distant star in the Milky Way Galaxy that looks very inviting and we think it may support life. The planet appears to have lots of water, beaches, video games and Little Debbie Cakes. We propose to embark on an epic journey, a mission to visit this tiny planet to see what's going on and what forms of primitive life, if any, might have evolved in the planet's environment. Should we find any life forms we would capture them and conduct experiments on them to learn what makes them so inferior to us. There are just one or two minor drawbacks to our plan. First, it will take us one billion years to get there in our fastest space ship. Second, the mission will cost a zillion dollars. But don't worry, the effort will be worth it. We'll learn all kinds of new things in preparing for the mission and when we do get to the new planet we may learn some amusing anecdotes from the moron life forms we believe might inhabit the place. Oh, and best of all, we can bring back some samples of rocks and dirt when we return home two billion years from now."

"I think you have an idea for a new sitcom John." Laughed Sam. "Seriously though, it would be quite unlikely for visitors from another world to even find earth. Even if they did discover earth the logistics and technology required to mount a project to visit, let alone come here repeatedly, is daunting to say the least even for a civilization that was far more evolved and advanced than life on earth."

"Some may argue that more advanced civilizations must exist, simply by pure statistical chance. OK, if that's true, it directly flies in the face of modern religious beliefs that man was specially and uniquely created by God. Why would an almighty creator make a sub-standard version of his greatest creation and place it on planet earth only to then have a far superior version make repeated secretive visits?"

Sam continued, "Don't get me wrong, I think the alien and UFO mania is a fascinating phenomena. Some people actually believe they have been abducted by aliens, taken aboard alien ships and studied. Numerous curious occurrences such as crop rings in Europe and so called Nazca lines in South American are attributed to aliens. Why do people look toward the heavens for signs of intelligent life or something greater than themselves when an unimaginable and sobering creation is everywhere, right now, and available to anyone who wishes to see?"

"It sounds like your not talking about UFO's any more Sam?" Said John questioningly.

"Yeah." Said Sam, sliding his hand up under the front of his backwards baseball cap scratching the top of his head for a moment. Repositioning his cap, Sam continued. "I think it's sad that most people know so little about the exquisite majesty of the creation of which, they a part. The cosmos, cells, DNA, the miracle of life itself are all there for everyone to know. It's not as if there is not good information about these things available. Its just that most people don't learn much science in school when they are young and then later in life they begin to believe that science is too complex for them to understand or, worse, is corrupt and is perpetuating a hoax on the rest of the world. Others believe that the only thing necessary to know is that their God made everything. Still others take the extreme opposite view that there is no creator and that everything just happened. So begins the great debate as people take up their respective sides on topics like intelligent design, creationism, Darwinism, fundamentalism, UFO's, alien abductions and the Lock Ness monster."

Sam continued, turning more somber but smiling while shaking his head in mock disbelief. "Did you know that there is a fundamentalist movement based largely in the United States called Young Earth Creationism. Its followers believe that the entire universe is only about six thousand years old and that fossils and light from distant galaxies were planted by God as,,,,well, basically as a joke on humans to make creation seem older and bigger than it really is! This is absurd and is beyond any logic. Even using the Young Earth Creationist central premise itself, that God is so magnificent he created everything in only six days just as the Bible says, these beliefs make no sense."

John could see a curious look of amused frustration appear on Sam's face as his demeanor changed, his eyes narrowing and for the first time small wrinkles appeared at the corner of his light blue eyes.

Sam held up one finger for emphasis and said, "Why would such a magnificent and powerful God make something as puny, limited and young as a six thousand year old version of creation? Isn't an ageless, unimaginably gigantic universe far more compelling for the existence of a powerful creator than something only six thousand years old? Is God truly king of all creation or just a cosmic jokester? Why would God go to the trouble to create a tiny, young universe then populate his special planet earth with his most precious creation only to have the last laugh by making distant galaxies seem further away than they really are and creation appear older than it really is? Was the creator just experimenting or maybe doing a dry run before the real thing? Why was the creator in such a hurry? Did the all-powerful and eternal God decide one Monday morning six thousand years ago to work an extra long week and whip up a tiny universe complete with misleading fossils buried in the ground and fake light from distant stars?"

"But, also, do you know what really gets me going?" Sam said as he shook his head once again, "What really gets me is that the God of modern day religions has limitations invented by his believers. Their God is nothing more than a super human being. An embodiment of what humans should strive to become. A tiny trinket that can be worn around ones neck and must be protected from science and non-believers. Religious extremists view scientific theories and hypotheses designed to uncover and understand the mysteries of the universe and creation as a conspiracy, a threat to their God. A threat! Come on now. Does an all powerful and all knowing God need protection from science?"

"In reality, the scientific process of hypothesizing, testing, and revising original hypotheses, reveals fascinating information about creation which only makes life even more magnificent. The fact is, scientific discoveries serve to further humble man before the creator. Nevertheless, radical fundamentalists of all religions believe that science is fostering a great atheistic conspiracy intent on questioning the existence, power and authority of their God. Quite frankly, I don't think that the creator of a universe containing billions of galaxies and a hole that is a billion light years across needs to be protected from scientists, non-believers, infidels, heretics, or blasphemers on this tiniest speck of cosmic dust called earth!"

John looked at Sam and would have sworn that Sam looked almost offended as he continued in a quiet but firm voice, "It's sad that allegiance to a tiny man-made God whose honor must be defended, sometimes to the death, is perpetuated by people who understand very little about his creation. Using religious belief and the strict adherence to so-called inspired text written by man as valid reasons to dismiss a deeper understanding of creation is a form of self-imposed ignorance. Doesn't it make sense that if your passion was to know and understand Mozart you would listen to his greatest symphonies and study them in detail? You would learn every note of his compositions, how every musical score and orchestration was developed, written and merged into a complete work. If you wanted to truly know the great artist Monet, you would study his paintings from a distance and up close. You'd come to understand his use of color, the genius he used to apply paint to the canvas in small sections assembling beautiful flowers and lush landscapes. The way his brush strokes were used to create variations of light and dark. You would examine every lily pad in detail."

Sam paused and, with piercing blue eyes looked directly into the setting sun in the western sky without squinting. "If you want to know the creator, you must study creation." Sam now sat up erect and proud. Breaking into a smile finally and with great relish Sam inhaled deeply as if he were tasting every molecule of the crisp evening air as it filled his lungs to capacity.

CHAPTER 9
John and Sam Move On

Sam slowly exhaled then turned and smiled at John who sat in stunned silence as he stared at Sam. John had been listening to Sam for what seemed to be the better part of the two hours and found himself hanging on every word. All of John's preconceived notions of Sam as some kind of peculiar child savant science nerd had melted away like an early spring snowfall in Virginia. John felt as if he was in the presence of a great and wise mind. The breadth of Sam's knowledge about a seemingly endless volume of topics from astrophysics to religion coupled with a fiery passion for creation itself, shook John's logic as he struggled to assimilate all he was hearing, and from whom. Since arriving at the rest stop his entire experience had been surreal and the fact that John now found himself hanging on the ageless wisdom of a boy, a child really, had John shaking his head in disbelief and questioning if he was not dreaming.

As the two sat in silence John turned from the sunset to look at Sam and for the first time began to notice some rather peculiar things about the boy. Sam's light blue eyes glistened and were ablaze as they reflected the vivid colors of the sunset. Sam rarely blinked, in fact, now that he thought of it John had yet to see Sam blink at all. Unruly, wispy light blond hair stuck out from under Sam's cap falling over his forehead. Sam's hair was as fine as a baby's, the individual strands of hair moved lightly in a nearly imperceptible breeze that blew ever so gently across the landscape from the southwest. Although Sam's light skin was flawlessly smooth, John noticed for the first time fine lines at the corners of Sam's eyes and around his mouth giving Sam's otherwise youthful face a distinctively older appearance. Sam's clothes were well worn and unremarkable other than seeming to be several sizes too large for his slight frame, which John surmised was the style for most kids Sam's age. But it was Sam's hands that now caught John's greatest attention. Sam's hands were those of an old man. Thick, stout fingers protruded from small beefy hands, palms thickened by calluses suggesting years of manual labor.

The backs of Sam's hands were cracked, dry and wrinkled, and were decorated with large veins and dark age spots. Sam's finger nails, though neatly trimmed, were rimed with rough, cracked cuticles. The nail on the thumb of Sam's left hand was blackened and looked as if it may have been smashed by a hammer weeks ago.

John's gaze shifted back and forth from the face to the hands of this curious, alarmingly intelligent boy when all at once he was snapped out of his trance as Sam jumped down from the table onto the ground and in a fluid, graceful movement turned and faced John. With his back to the western sky Sam's head was perfectly framed by the setting sun, making him appear as if he wore a crown of light.

"Well John, I really should be going and you must want to get back on the road and continue your journey." Sam said quite matter of factly and with no fanfare or forewarning.

"Well yes." John remarked as he straightened up. Still seated on the table and squinting back at Sam with a puzzled look on his face, John suddenly realized he could not remember what he was just thinking of only moments earlier. "I probably should get going too." John said.

Sam smiled as he turned lazily on his heal toward the setting sun. He took a slight childlike skip for his first step and began to head up the slight incline toward the parking lot at a brisk pace. As he walked away Sam turned and glanced over his shoulder toward John and with a wave of his hand Sam said, "Its been great talking with you John. Thanks for letting me go on and on about stars, light years, holes in space, UFO's and stuff. I don't find many grownups willing to listen to me. Most old people think its weird having a kid talk to them about things they probably never really think about very much. Drive safe ok. Maybe we'll bump into one another again."

John got down off the table, planted his feet firmly on the earth, stretched his back and began to follow Sam up the slight rise to the parking lot while calling out to Sam, "Hey, wait a minute."

Sam paused and turned again toward John, looking at him with those piercing blue eyes and said as he continued to walk away from John, "Enjoy the sunset John, and remember, some of the stars that you'll see later tonight when the sun goes down are actually galaxies containing billions of stars. The light you'll see left those galaxies billions of years ago."

John shaded his eyes with his hand and squinted as he looked at the sunset again. He was stunned to see that it had seemed not really to have changed at all since he first sat down with Sam. It was as if the time he and Sam had spent together had passed in but a moment. While it was just dusk, the sun was still visible and sinking behind the tree studded hill as it had appeared when John and Sam began talking. By now, Sam was up the stairs and across the parking lot heading over a small rise and into the forest directly toward the setting sun.

John took a few quick jogging strides toward Sam's black silhouette as it began to merge with trees, now simply blackened vertical shadows lining the edge of the woods. John called out, "Sam, I....I..., Thanks. See ya later." He saw Sam hold up his hand and wave one last time without turning around. A casual friendly wave of his right hand raised slightly above Sam's head was the last thing John saw before Sam simply vanished into the shadows of the wooded area at the edge of the parking lot.

Not really knowing what to think about anything at the moment John took a deep breath and mumbled to himself, "That was weird.... Interesting, but really weird". He looked at his watch and saw that it was about 6:15 PM. He began to walk slowly toward his car. As he walked he thought about all that he and Sam had talked about. With each step, John felt more invigorated as he walked toward the parking area thinking about the vastness of the universe and the way Sam had described the many wonders that lay beyond John's imagination. He tried in his minds eye to picture traveling at many times the speed of light to the furthest reaches of the cosmos, which he now knew to be feasible only in one's mind. There he found nothing but a countless number of new galaxies and more empty space. Infinity was hard to imagine. He thought about the hole in the universe and wondered why it was there. Did it have a purpose or was it just a random speck in a gigantic creation that stretched beyond imagination in all directions? Was there some other alien creature on some unimaginably distant planet thinking the same thing at this very instant? Or better yet, were there an infinite John Mitchells inhabiting billions of parallel universes walking to their cars, mirroring every movement John made and every thought that popped into John's head countless of times in a countless number of identical parallel worlds?

John stopped abruptly. For the first time in his life that he could recall he was overwhelmed with amazement and awe that surpassed even the birth of his two girls. It was sobering to realize that while he was only the most minute part of a creation that stretched beyond imagination, he could literally travel endlessly through the cosmos in his minds eye. The limitations of the physical world such as the speed of light, time and distance presented no barrier for the human mind. John wondered why in a creation that was billions of years old, that the very moment in which he found himself was his and his alone. In this moment at a rest stop in Virginia with the sun slowly setting in the western sky, he, John Mitchell in a single instant could escape to a galaxy thousands of billions of light years away from earth. There, he could hover near one of the galaxy's hundred billion stars and observe its planets and their moons. Perhaps one of these might harbor life, fragile and wonderful. Then in a heartbeat John could travel back to Virginia and wonder what role he played in an ageless and limitless creation that was as real as the earth he could feel beneath his feet. John was indeed, at that moment, filled with a reverent sense of awe, wonder and joy. He was part of something infinitely larger and far more grand than just himself. John's inner voice thirsted to understand how he was connected to a gigantic hole somewhere in the furtherest reaches of the cosmos and hundreds of billions of galaxies, some of which, would soon pepper the night sky with light that was as old as creation itself. John found himself wanting to know more.

CHAPTER 10

Neva

Approaching his BMW, John pressed the "door" button on his key fob though he was oblivious to the familiar beep and solid click as the doors unlocked and the indoor cabin lights illuminated. John's mind continued to wandered through the vast regions of a newly revealed universe. Robotically, John got behind the wheel, started the engine, fastened his seatbelt and backed carefully out of his parking space even though the rest stop remained deserted. Blinking hard and shaking his head to refocus his thoughts on driving but, purely out of driving habit, John looked briefly in his rearview mirror as he shifted from reverse into drive and began to slowly accelerate forward, straightening his wheels and pointing the BMW toward the ramp that returned drivers from the rest stop to the highway. In his rearview mirror and barely visible in the dimming light John noticed a single car coming slowly into the rest stop parking area. Even at some distance John observed that the approaching car was listing noticeably to the passenger's side. John stopped momentarily looking more carefully in his rear view mirror as the lone car lumbered slowly into the parking area and crawled to a stop several spaces down from where John had been parked only moments earlier. In no particular hurry and somewhat curious that there was now another car in the otherwise deserted rest area John twisted in his seat to get a better look at the parked car.

The car was a large green American made sedan sporting a cloth hardtop roof. A few bumps and dents were visible along the vehicle's right side. Otherwise, the car appeared to be in pretty good shape considering that by John's estimate the car was probably of an early 80's vintage, large and boxy. John could see instantly that the reason the car listed to the passenger's side was that the right front tire was completely deflated and nearly shredded. Bits of mangled rubber barely clung stubbornly to the metal rim. The driver obviously had gotten the flat on the highway and rather than risk pulling off to the side of the interstate, the car's driver had driven to the rest stop to change the tire. The damaged tire was beyond repair.

John noticed that the driver was a woman and he continued to watch as she turned off the ignition and open the driver's side door to get out. Closing the door behind her the woman looked first at the left front of her car then walked around to the back of her car looking somewhat puzzled as she searched for something amiss. Rounding the back of the car she walked slowly along the passenger's side and finally stopped as she noticed the flat right front tire. Visibly heaving a long sigh the woman stared up toward the heavens and put her hands on her hips. Taking another breath, the woman looked down once again at the ruined tire and shook her head.

Except for John and the woman, the rest area remained completely deserted. John had no choice but to return to his parking spot and ask the woman if he could help. Besides, it was the chivalrous thing to do.

"Looks like you have a flat tire." John said as he got out of his car and walked slowly toward the woman. "Did it just happen?" He added, smiling in a purposefully friendly manner while moving cautiously so as to not give the woman any cause for alarm given that they were the only two people in the rest area.

"There was some debris on the road a few miles back that I couldn't avoid and then a short while later I noticed that the car began to steer funny. I started to slow down and fortunately was within site of the rest stop so I pulled off here. I'm glad I didn't have to stop along the side of the road." She said, then adding "With the construction there was hardly any room to pull over and besides it's getting dark and traffic goes so fast on this highway".

"Well you are a lot safer here than along the side of the road." John agreed. "Perhaps I could help and change the tire for you?"

"Oh that would be so kind, but I'd not want to impose. You probably need to be on your way." Said the woman. Then with another sigh she continued "But, to tell you the truth, I wouldn't even know where to begin to change a tire. I'm surprised there isn't anyone else here. I'd be very grateful to you if you could help me." She said while looking around as if she just became aware that there was no one else at the rest stop.

"Not a problem." Said John, continuing to flash his most friendly smile as he walked toward the woman and held out his hand. "I'm John, John Mitchell. I would be happy to help you. Besides, it shouldn't take but a few minutes to change the tire and have us both on our way."

John guessed that the woman was in her mid to late 50's. She wore a neat, but dated, flower print dress that came to her knees. Her light waist-length zippered sweater jacket with a hood was open but still provided a little warmth against the cool of the early evening. Her shoes were brown and comfortable looking with hardly any heel. The woman had grey wavy hair cut about chin length which neatly frame a slightly round and extremely pleasant face. Just a little over 5 feet tall, the woman was not heavy but full figured and rather fit looking.

"I'm Neva." The woman said, smiling as she reached to take John's hand. Fading light from a setting sun was just enough to highlight beautiful flawless dark olive skin and deep brown eyes that complimented an engaging smile and pearl-white, perfect teeth. "Nice to meet you John, and I really do appreciate your offer to help." She added, extending her right hand toward John. John grasped her hand which felt small in his. The skin of her hand was dry, but smooth and warm. Her handshake was firm and confident and John increased his grip slightly in response. Neva looked John directly in the eyes when they shook hands, holding his gaze as they greeted. Neva continued to smile warmly and gave a little mock curtsey acknowledging non-verbally that she was delighted John had offered to come to her rescue.

"So this is a Buick right? What year is it?" John asked, tilting his head toward Neva's car as he and Neva stopped shaking hands and began to walk toward the back of the car to the trunk. "Looks to be in pretty good shape. I wonder if this has a full sized spare or a donut?" John thought aloud.

"A donut?" Neva asked, sounding perplexed.

"Yes, you know, a small sized spare tire? A lot of cars don't have full size spare tires any more. To help reduce weight, increase trunk space and improve gas mileage most car makers started to provide a smaller spare tire in their cars. They're called donuts. It's smaller than a regular tire and is not to be run at high speeds or for long distances. It's just to help you limp along until you can get the regular tire fixed or replaced. In fact, you're not supposed to drive over fifty miles per hour when you have a donut on your car and generally, they are only good for few thousand miles." John explained to Neva, providing far more information than Neva probably wanted or needed to know .

"Oh my." Said Neva suddenly becoming concerned as she began searching for the trunk key on her key ring. For some reason General Motors for the longest time was one of the only automakers who believed that every car should have two keys; a square one for the ignition and a round one for the doors and trunk. John smiled to himself as he noticed that just like Lucy, Neva's key ring contained at least twenty different keys probably two-thirds of them for which she'd forgotten what they were for. As a result, it took Neva a moment or two to find the round key to the trunk.

"Ah, I think this is the one." Said Neva proudly holding up the round key for the trunk. "I couldn't tell you what type of spare tire this car has since I've never looked. I rarely even open the trunk to tell you the truth, I just toss what ever I need into the back seat where its easy to reach. I think I got this car in 1980 or 1981."

"Wow, that's a classic!" Exclaimed John as Neva put the key into the trunk lock. Neva turned the key. The hinges creaked as she opened the trunk's lid. Just as she said, the trunk looked as if it had never been used. Only a thin mat made of black cardboard and a few dried leaves occupied the voluminous trunk. John thought to himself that he hoped the spare had air in it. If Neva had never checked the spare tire the chances were that spare could be partially deflated and not usable given the car's age.

John lifted the cardboard mat to reveal a jack, tire iron and a full sized spare tire. The spare looked to have never been used. Unfortunately, just as John had feared the black rubber of the tire was cracked and faded. The edge of the tire easily gave to his weight when he put his palm on its rim and pushed down.

"I'm afraid we have a problem." Said John straightening up and looking from Neva to the spare tire. "Your spare looks to be flat. It probably just lost air over time in the trunk so we'll not be able to use it. Looks like you are stuck here until we can get a tow truck to come. The tire on the car can't be repaired so you'll probably have to get a new tire as well once you get to where you are going." John offered to call for help.

"That would be very nice of you." Neva said with a grateful smile. "I don't have a cell phone which is kind of silly but I guess I'm just kind of old fashioned. I'm so sorry to be such trouble and hold you up like this. I can't believe I never checked the spare tire. I guess I never really ever thought about getting a flat."

John smiled at Neva letting her know that it was no trouble as pulled out his cell phone and dialed 911 since he had no idea how to reach a towing service and he was not a member of AAA. When the 911 dispatcher answered John explained the situation and asked if they could provide a number of a tow service. Based on her accent and the timbre of her voice the dispatcher sounded to be a middle age female from the deep south. She wasn't very happy and made it clear to John that he shouldn't use 911 unless it was a true emergency. Nevertheless, following her mini-lecture on the proper use of 911 she quickly gave John a number and hung up without repeating the number or asking if John got it. Fortunately, since John was quick with numbers it was easy for him to remember the phone number which he dialed immediately after the crotchety dispatcher abruptly hung up.

The phone rang twice before a man's voice answered, "Hello, Fred's Tow. How can I help you?"

While John wasn't quite sure what mile marker they were near, he explained his location and was pleased to hear that Fred's Tow was familiar with that section of I-95 and knew exactly which rest stop they were at. Unfortunately for John and Neva, Fred's Tow said it would take at least an hour to get someone out there. They only had 2 trucks and one was already on I-95 helping to clean up an accident about 15 miles north of John's and Neva's location. The other truck was in rural Virginia horse country towing a Lexus to the Lexus dealer in Fredericksburg, VA. Turns out the Lexus, it appeared, had developed a computer fault that affected the entire electronic system and the car was stone dead. This time it was John who was the recipient of more information than he needed to know but he thanked Fred's Tow and asked if there were any other services in the area that might be able to help. The man on the other end of the line said that he was actually the closest to where they were and that most other tow services would take at least as long to get there as Fred's even if the others had trucks available. John thanked the man on the other end of the line said they would be waiting at the rest stop for the first available tow truck from Fred's Tow. John gave Fred's Tow his cell phone number and asked that they call him if their ETA changed.

John pressed the end-call button and explained the situation to Neva who was not surprised as she could tell from hearing John's side of the conversation as well as the look on John's face that help was not going to be coming quickly.

"I'm going to stay here with you Neva until the tow truck comes." Said John. I'm not in a big hurry and I want to make sure you get this taken care of. Besides, there's no one else here and to tell you the truth I wouldn't mind hanging around an seeing the end of that beautiful sunset." John nodded his head in the direction of the deepening and varied colors that were forming in the western sky.

"Oh I simply couldn't ask you to do that. I just feel awful about this." Said Neva apologetically. "I'm sure that other drivers will come along from time to time and I don't mind waiting in my car until the tow truck arrives. Really its very kind of you but I'll be fine. Thanks for all you have done, but please, you should be on your way. You must be anxious to get to wherever you are going."

"Nope." Said John. "I insist. I'll not leave a lady in distress. And as I said, it's a beautiful evening and I'm in no hurry. How about I get us something to drink from one of the vending machines and we can just sit and talk until Fred's Tow gets here."

Neva smiled and closed the lid to the truck. "That's very kind John. Most people wouldn't have even stayed to see if they could help in the first place. I really do appreciate this. But if you insist on staying I insist you let me buy the drinks while we wait."

"Fair enough." John said and gave Neva a slight bow as if he were a royal knight who served at the pleasure of Neva the queen. Though it was unseasonably warm for March the air was beginning to cool just a bit although it was certainly not uncomfortable by any measure. John and Neva walked down the steps and headed toward the restrooms and vending machines.

CHAPTER 11

Origins

As they walked to the vending machines John and Neva agreed they would take their drinks to a picnic table to watch the sun set. If it became too chilly they would return to one of their cars. Making small talk while they got their drinks John told Neva a little about his family and said he was on his way home to Raleigh from a business trip in Washington. Neva told John that she was a professor of Biological and Ecological Engineering from Oregon State University. She was on two-year sabbatical and was writing a book on the Chesapeake Bay watershed which was why she was in Virginia. Neva briefly explained to John that the Chesapeake watershed was massive and encompassed the entirety of the states of Maryland and Delaware, half of Pennsylvania, about one-fifth of New York and a good two thirds of Virginia. The entire system was under tremendous environmental pressure from chemical fertilizer runoff and she was examining the potential impact that the loss of a number of species would have on the region's ecosystem. Neva had no family to speak of and she enjoyed traveling in her Buick and doing research for her book.

Both John and Neva chose bottled water from the vending machine. True to her word, Neva bought and with drinks in hand she and John moved to the picnic area and sat down at the same table John and Sam had used earlier in the day. Unlike Sam and John, however, John and now Neva chose to sit on the actual bench seat opposite the setting sun so that they could see the final remnants of the sunset unfold in front of them. John removed the cap from his water, took a sip and was about ask Neva more about herself when Neva spoke.

"Isn't it beautiful?" Neva began, seemingly just talking to herself as if John was not there and she was simply articulating her thoughts to no one in particular.

"In all creation" Neva continued, "I think that a sunset is one of the most beautiful things anyone can see. And to think that while we are watching the sun set and our day is coming to an end, someplace on the other side of the world the same sun is rising and a new day is beginning. A continuous, endless sunrise and sunset somewhere around the earth that links every creature on the planet. Its been going on like that for billions of years. It's simply amazing."

Neva paused, breathed deeply, closed her eyes and inclined her face slightly upwards. Neva smiled as she enjoyed the light and warmth from the lengthening rays of the setting sun. The setting sun's light lingered as it carved ever changing patterns of illumination through the trees of the forest beyond the parking area. Neva's face took on a radiance that rivaled the sun itself and John noticed that Neva looked, quite simply, beautiful.

For the second time today John found himself sitting in stunned silence staring at someone he'd just met, someone who had immediately captured his very thoughts. Recalling his conversation with Sam, John said out loud without any intention at all, "This is unbelievable."

"Oh sorry? What's that?" Asked Neva without turning her face away from the west, although she suddenly seemed to be aware once again of John's presence beside her.

"You might find this hard to believe…" Began John. "… seriously, not thirty minutes ago, just before you limped into the rest stop, I had just finished the most amazing conversation with a boy sitting at this very bench. We talked about the sunset, light years, the age of the universe….all kinds of things. It's as if….I can't really explain it…but, its like you just picked up where Sam and I left off!" John paused, then asked, "Who are you?" Turning toward Neva, John shook his head and squinted his eyes tightly closed while shaking his head quickly as if to try and clear a mental fog.

"Why would I find it hard to believe that you were talking to someone else about the sunset?" Neva said while turning her face slowly, reluctantly, away from the sun and opening her eyes to meet John's. Neva continued, "It's not unusual at all to get philosophical while looking at a sunset you know. It's actually very natural. Its also not uncommon at all to think about where all this came from and why we are here. It's an ability that humans have that is unique among all living creatures. Of all life on earth, only humans have the luxury of being able to spend time gazing at a sunset as they think about creation and wonder how and why they are here. Animals can't do that. Animals need to spend all of their time resting, mating, eating or trying to avoid being eaten! I don't find it odd at all that you might have sat here with someone and had a conversation about life and the universe. Its probably the most human thing anyone can do."

John thought for a minute and then said, "I guess that makes sense. Its just that I pulled into this deserted rest stop and meet Sam, this boy who couldn't have been more than ten or eleven years old. Sam knows all kinds of things about about galaxies, parallel universes, quantum entanglement mumbo jumbo and giant holes in the cosmos. For a boy, Sam was really, really smart. He made me feel like I know nothing….which, turns out I don't!" John said shaking his head with a little chuckle.

"I'm no scientist, but Sam helped me understand how big and ancient the universe actually is. It all made so much sense the way Sam explained things. He was just a kid and he made me think about things differently than I'd ever done before. Now, out of nowhere I meet you and you start waxing philosophical about the rising and setting sun, all the creatures on the earth, and…"

John paused for a moment to look at the sun set then continued, "Its all just a bit weird that's all. But at the same time I have to tell you….a Darden graduate whose worked in venture capital for the past ten years is suddenly fascinated by all this. All the more surprising given that fact that for the last twenty-five years I haven't even given a moments thought about anything that has to do with the universe or creation. That is, other than to watch Animal Planet with my daughters once in a while."

Neva smiled and, offering to change the subject she asked, "What are your daughters' names John and how old are they?"

"Naomi and Jessica" said John. "Naomi is seven and Jessica is five. They are awesome and both are beautiful like their mother. I am so lucky".

"Those are lovely names. From the pride in your eyes when you talk about them, I'll bet they have their dad wrapped around their little fingers." Laughed Neva.

John laughed as well saying, "That's for sure. There's nothing I wouldn't do for them. Unfortunately, they know it!"

It was John who steered the conversation back to his newly revealed insights, "I guess I've been thinking about the universe since talking with Sam. I'm simply amazed that in something as big as the cosmos and, with the earth being just the tiniest speck of nothing in the midst of an endless creation, I have two wonderful daughters and an amazing wife. My whole universe revolves around the three of them. We watch Animal Planet and see all these creatures: bugs and lizards and fish and plants and gorillas and such. It makes me wonder how all this came to be. Why on this microscopic speck of dust in a universe that is almost beyond imagination has life of all kinds appeared? And with all this going on, I end up with two little girls who themselves are pure works of art. It really boggles my mind."

"Now who's getting philosophical" Quipped Neva with a laugh. She paused and gave John a thoughtful look before going on. "You have hit on the thing that makes humans human. The fact that you even spend a single moment of your life thinking about such things is otherwise unheard of in creation. All other creatures must spend every moment trying either to survive or reproduce. Even if they had the brain power no other creature has the time to get philosophical like humans do."

"And you know what else is uniquely human." Neva continued.

John shook his head "No, what?"

"Your stopping to help me. You even went out of your way to offer help when there were a hundred other things you could have been doing. Some believe that these uniquely human traits are ingrained in our genes and are the result of evolution of the human species. Others argue that when you understand how genes function and cells work it's hard to imagine that all of this, from a lowly bug to John Mitchell who is kind enough to stop to help a lady in distress, came about purely by random chance."

John spoke, "Sam and I talked a little about that earlier. He said that there is a big difference between the origin of life and the evolution of life. These are two different things and they often get confused and co-mingled in all the debate about whether or not man was created or evolved. Personally, I'm very confused. I'll tell you one thing though, I don't see how someone as intelligent as you or Sam"…John smiled. "…or, as beautiful as my girls could have evolved from some primordial soup full of chemicals no matter how many billions of years the oceans sloshed around."

"Ahhhh…" Neva grinned and a sudden burst of enthusiasm became apparent in her voice. "… organic molecules, molecular biology, cell biology, developmental biology, evolutionary biology, Darwinism, intelligent design, creationism…all fascinating subjects John."

"Who in the world ARE you people?" John asked again with astonishment. "I keep running into people out here in the middle of nowhere who know so much about things I've never even thought about until now! I don't think I'm dreaming but if I am, this is a strange one to be sure."

Neva smiled, placed her hand on John's arm giving it a gentle, friendly squeeze and said, "I guess it's just your lucky day to run into some people who have answers to your questions John. You've happened to bring up some of my favorite topics. If you're really interested we can continue your dream and talk a little about how all this came to be." Neva gestured with her palms uplifted in the universal sign for "everything around us". "It really is simply fascinating to know something about how life works. A little knowledge can't hurt. It might really open your mind you know."

"Why not." John said, shrugging while returning Neva's smile. "If you are half as interesting as Sam I'd love to hear some of what you know. To tell you the truth I haven't thought about the universe or creation for as long as I can remember. For whatever reason today just seems to be my day to learn everything that I should have learned in school had I been paying attention! Besides, just wait till I get home and tell Lucy and my daughters about you and Sam. From what little I know to this point chances are after talking with you I'll be a virtual fountain of information, able to dazzle anyone with snappy dinner conversation. I'm all ears professor. The podium is yours!"

CHAPTER 12

All Things Original Matter So

"Well, to begin with…" Neva said, "… I'm sure you realize, don't you, that everything, and I mean everything, is made of atoms?"

John nodded, "Yes, I know about atoms. They are made up of protons, neutrons and electrons. The protons and neutrons are in the nucleus and electrons circle around the nucleus. Kind of like the planets orbiting around the sun."

Neva smiled obviously impressed. "Well, its actually a bit more complicated than that but basically you have it right. You obviously paid attention to something in school. Do you remember that the protons have a positive charge, the neutrons are neutral and the electrons have a negative charge?"

John nodded acknowledgement proudly adding. "I sure do remember now that you remind me. It was Ms. Sullivan, my teacher in third or fourth grade as I recall. She was one of my favorites. I remember building a model of an atom in her class. I can still see it in my mind today. It was a thing of beauty I'll tell you." John concluded with a smile.

Neva continued, "Well, what you probably did not learn in school is that there are special forces in the atom call the strong and weak nuclear forces that work to keep all the parts that make up atoms bound together. Furthermore, we now know that protons and neutrons are made of even smaller things call quarks and gluons. Each proton and neutron contains three quarks which are held together by gluons."

"I didn't know that. Quarks and gluons must be really tiny." Said John.

"Oh very tiny." Said Neva. "To try and imagine how tiny, think about an atom that is the size of the earth. That is to say that the outer-most electrons of the atom, like the model you built in Ms. Sullivan's class, are orbiting at the surface of the earth and the nucleus is at the very center of the earth. On this scale a proton would be about the size of a football field. Even tinier yet, however, quarks, gluons and the electrons themselves would be only the size of a golf ball. With this size comparison you can see that atoms seem to be mostly empty space…a lot like the cosmos in fact. Interestingly too, neither electrons, gluons or quarks have any mass which is to say that they don't weigh anything. And yet, everything from the most gigantic stars to a tiny speck of dust reflected in a ray of sunshine streaming through a window is made up of atoms. And atoms, just like the universe, are essentially all the same thing…empty space! So, there you have it. What makes up everything in the universe doesn't …."

"Wait a minute." John interrupted, stopping Neva in mid-sentence. "Back up just a moment because, I'm confused. You just said that electrons, gluons and quarks don't weigh anything and that atoms are mostly empty space. If electrons don't weigh anything and protons and neutrons are made up of things that don't weigh anything and most of an atom is empty space, how can atoms weigh anything? I also remember a little about the periodic table and know that each different atom has a mass that increases as you move from light things like hydrogen or helium to the heavy things like iron, gold, and lead. If everything is made up of atoms and everything has weight then atoms must weigh something?"

John stopped and gave Neva a smug grin believing that he had caught Neva on a technicality that she was just trying to gloss over. After all, John made his living challenging people on the details of their claims and he prided himself on not letting a little factoid slip by that was critical to an argument.

"Ah ha!" Neva exclaimed excitedly. Rather than looking like a deer caught in the headlights which was the usual reaction of someone when John caught them on a technicality, Neva looked at John with a proud twinkle in her eye like a teacher who was thrilled to see the light bulb go off in her prized student's head. "You've hit upon a very, very important question." Neva continued. "You are absolutely correct. How could an atom weigh anything if it is made up of mostly empty space and particles that don't weigh anything? Quite puzzling to be sure and it is something that has had scientists scratching their heads for quite a long time actually. The good news, however, is that just within the past couple of years the answer to this prickly question has been discovered. What scientists have learned is that what we perceive as empty space is not actually empty space at all. The holy grail of modern particle physics has been to determine what, if anything, makes up the empty space in atoms. It had been hypothesized that something must exist in all this nothingness since without something there it is impossible to explain why elementary particles such as the electron, quark and gluon, which alone have no mass themselves, can form atoms that do have mass. In other words, without something filling this empty space, atoms indeed would have no mass since electrons, gluons and quarks have no mass. As a result, a number of years ago physicists hypothesized that all the empty space in atoms was actually filled with something called the Higgs field. The Higgs field, in turn, is composed of something called Higgs particles or as they are more commonly known, the Higgs boson. It is the movement of electrons, gluons and quarks through this Higgs field which is believed to give atoms their mass. It's something like swinging your arm around your body under water in a swimming pool. When you do this out of the water your arm feels relatively light. On the other hand, when you do this under water your arm feels heavier because of the resistance of the water. In the same way, electrons, gluons and quarks attain mass as they travel through the Higgs field at the speed of light. As these elemental particles move through the Higgs field, they encounter Higgs bosons and become slowed down due to the resistance, gaining mass as a result. Just like your arm feels heavier and moves more slowly through water than through air, electrons, gluons and quarks slow down just a bit and become heavier as they move through the Higgs field compared to how they would behave if the were in fact moving through true nothingness. The hypothesis and ultimate discovery of the Higgs boson was so important to the understanding of physics and how our universe works that the Higgs Boson has been nicknamed the 'God Particle'. The Higgs boson is fundamental to everything since in effect, the God Particle is what gives everything substance."

"One of the truly incredible things about creation is that there is a logical, comforting and beautiful universality that binds everything together. Just think for a moment of atoms and the building blocks of atoms as mostly empty space composed of God Particles. Since everything is made of atoms this means that at the most basic level you, this picnic table we are sitting on, our cars, the trees, the air we breath, the oceans, every rock, animal, planet, star, solar system, and galaxy all share something in common that is so fundamental, so empty and yet so full of an elusive, mysterious nothingness. Everything therefore, is ultimately the same. Now that's really something isn't it?"

John thought for a moment then commented, "You know, I really felt small when Sam was talking to me about the size of the universe and the hundreds of billions of galaxies it contains. And when I was thinking that everything is mostly empty space, my self esteem fell even lower. But, now you tell me that along with everything around me I'm actually not empty space but, in fact, I'm mostly something and not empty space at all. I'm filled with God Particles! Suddenly I am feeling much better about myself. Thank goodness for the Higgs boson since without it I really would be next to nothing!"

"That's an interest way to look at things." said Neva. "But, to tell you the truth, it gets even better than that. Perhaps you'll feel even more important if we talk about where all these atoms and the space within them came from in the first place and how at the atomic level you are identical to the mightiest stars in the universe."

It was time for John to learn something about the Big Bang, particle physics, atoms and molecules.

The universe is believed to have formed about fourteen billion years ago. While no one can ever prove for certain how or when everything in the cosmos began all the evidence points to a violent and quite remarkable beginning. Known as the Big Bang, the theory is based on original observations made by the astronomer Edwin Hubble in the mid 1920's. Hubble made a series of highly precise and painstakingly complex measurements showing clearly that distant galaxies were moving further away from earth and from each other. In other words, Hubble concluded that the universe was expanding. Then in the early 1930's a Belgian physicist who also happened to be Roman catholic priest named Georges Lemaitre confirmed Hubble's findings and proposed the idea of the Big Bang. Lemaitre's findings were later confirmed and supported by others as the modern laws of physics became better understood. Simply put, the Big Bang theory argues that the cosmos we observe today began as single point of pure energy. In the beginning, there were no atoms, no electrons, no quarks, no gluons, no Higgs bosons, and neither time nor empty space existed. In the beginning there was in fact, literally and truly nothing since mass, time and empty space simply did not exist. That may be a difficult concept to get one's mind around but the laws of physics point clearly to the fact that in the beginning there was truly nothing except for a single point of pure energy. Then, for reasons completely unknown this single point of pure energy became unstable and exploded; i.e., the Big Bang. During the first one billion, billion, billionth's of a second after the Big Bang this immense amount of pure energy was converted into particles, specifically quarks, electrons, gluons, and Higgs bosons which themselves instantly combined to form protons, neutrons and electrons. These original protons, neutrons and electrons in turn combined forming unfathomable amounts of the smallest and most fundamental atoms that exist, namely hydrogen and helium gases. Therefore, in the first few moments of creation following the Big Bang literally all the elementary particles that comprise the entire universe that we know today were formed.

What happened next is what binds all things in the universe together in a most intimate fashion. As the universe continued to expand and cool after the Big Bang, unimaginably massive clouds of hydrogen and helium gasses billions of light years in diameter began to condense and form the first galaxies. These were galaxies, however, without stars or planets as these first galaxies only existed in space as gigantic collections of hydrogen and helium gasses. Within those first galaxies, localized and especially dense areas of hydrogen and helium were smashed together as a result of intense gravitational forces created by the sheer mass of all the hydrogen and helium atoms. The density of these local collections of hydrogen and helium gases increased even further as gravity relentlessly crushed more and more hydrogen and helium atoms together. As a result of this violent compression of atoms the temperature of these massive gas clouds began to rise and ultimately give birth to the first stars. It was in the cores of these first stars, where under unimaginable temperatures and pressures, the process of nuclear fusion began. Nuclear fusion is the combination of smaller atoms to form larger, heavier atoms. In the very hearts of these early stars, hydrogen and helium atoms combined with themselves and each other to form an abundance of heavier atoms such as carbon, nitrogen and oxygen which just happen to be the essential elements of all life.

However, even with all this nuclear fusion occurring within the cores of the first stars creation was not done yet. After hundreds of millions perhaps even a few billion years as these first stars burned out and collapsed upon themselves, temperatures and pressures rose even higher, forming even heavier atoms such as iron, magnesium, calcium and potassium. Eventually these first stars exploded and all of the different atoms that had been baked in the star's inner cores for millennia spewed outward into the star's local regions within their galaxies. These exploding stars are known as super novas. The deaths of these early stars eventually gave birth to new stars and the solar systems surrounding them. Those new second generation stars and the planets that orbit them were made of the atoms formed in the original parent star. Indeed, the sun is actually a second or perhaps even a third generation star which was born following the death of a parent star that exploded as a super nova some five billion years ago. That explosion gave rise to the sun and the planets that comprise our solar system as we know it today, including the planet earth.

"So.." John again interrupted Neva for a moment, "…the atoms that make up the earth, and probably me, were created at the time of the Big Bang and at one time were part of a star that existed even before the sun?"

"That's right John. Every atom in your body was once not even an atom. According to the Big Bang theory everything started as pure energy. The Big Bang explosion caused that energy to be converted into simple hydrogen and helium atoms. Those hydrogen and helium atoms were squashed together in stars to make carbon, oxygen and other atoms that make up all the different atoms in the universe. The protons, neutrons, electrons, quarks, gluons and Higgs bosons in every atom within you were actually formed fourteen billion years ago and have been cooked in stars for billions of years before the sun even existed. You and everything around you is not only fourteen billion years old but you are quite literally made out of stardust that was originally born from a tiny point of pure cosmic energy. Now, how does that make you feel?" Neva stopped and looked at John with the same pride that a mother has when looking at her newborn infant.

"Really pretty incredible. I guess I'm not as small or insignificant as I thought!" John said, taking a deep breath and puffing out his chest a little with pride.

"And there's more." continued Neva. "Atoms are the basic building blocks of everything. You remembered the periodic table from chemistry class? Each element or atom represented in the periodic table of the elements is different, not only in the number of protons, neutrons and electrons, but in size, weight, physical and chemical properties."

"Even though each of the more than one hundred atomic elements of the periodic table is made of the same things as every other element, every atomic element is physically and chemically unique by virtue of the different number of protons, neutrons and electrons from which it is composed. These differences not only give each atomic element their different properties but allow atoms to combine together forming virtually a limitless number of larger complex particles called molecules. Molecules may look and act nothing like the atoms they are made of. An atom of sodium, for example, is highly reactive and explodes when mixed with water. Atomic chlorine is a deadly poisonous gas. Yet, when sodium and chlorine combine chemically they form sodium chloride, or common table salt. Oxygen and hydrogen, two highly explosive gases combine to form water. So, the combination of highly toxic or dangerous atoms form things like salt and water molecules without which, life on earth would be impossible."

"As you might guess, the truly amazing thing about atoms and molecules is their ability to combine to form an endless variety of substances that give everything known its specific form and function. These combinations, and the way atoms and molecules interact with one another, follow very precise laws of chemistry and physics. These laws, or rules if you will, are quite well understood and the more that has been learned about the underlying principles that govern atomic and molecular interactions, the more fascinating creation becomes. All substances, everything, from the air we breath to the most complex organism or most massive star are made from atoms which are all made of the same identical and fundamental particles which are as old as creation itself."

Neva continued talking and John sat listening to every world as she began to talk about carbon, the atom of life on earth.

"Of all the atomic elements, one of the most interesting and certainly the most important to life on earth is carbon. Carbon is the atom upon which all life on earth is based. In physical appearance in its most pure form carbon is rather uninspiring, existing chiefly as a black powder. Basically soot. As you probably know, pure carbon when highly compressed under tremendous pressure forms coal. Compressed even further, carbon becomes a diamond. But, even a diamond pales in comparison to the many complex and diverse substances that carbon can form when combined with atoms of hydrogen, oxygen and nitrogen. The carbon atom is literally the backbone of life. Because of carbon's unique atomic structure, carbon can combine with itself to form molecular carbon rings or long molecular carbon chains. Carbon rings or chains can, in turn, combine with hydrogen, oxygen and nitrogen atoms in very specific ways to form sugars, alcohols, proteins, fats, vitamins, and DNA. Every living creature since the beginning of life on earth has been made out of molecules for which the primary component is carbon formed in the cores of stars born during the first few billions of years after the Big Bang. Remember, carbon, hydrogen, oxygen, nitrogen and all the other atomic elements were created from hydrogen and helium atoms formed during in the very first moments after the Big Bang. Through their common atomic structure every living creature is linked to all the other matter throughout the entire cosmos no matter how large or how small, how near, or how far. All things are part of the original creation because they are made from atoms that were formed at the very beginning nearly fourteen billion years ago."

A black ant hesitantly appeared from between two planks on top of the picnic table where John and Neva sat. Traveling on some unknown but vital mission to somewhere only it understood, the ant raced in starts and stops across the top of the picnic table. On its journey the ant halted often to inspect random specks of dirt or tiny crevasses in the wood that decorated the surface of the picnic table which was badly in need of painting.

Neva continued, "More than two thousand years ago some Greek philosophers and Roman poets imagined that the universe and everything in it was made of countless minute particles they called 'atomos' that had existed since the dawn of time. The Greeks and Romans even speculated that the universe came about purely by chance as a result of these 'atomos' crashing into one another by countless numbers. Turns out, these early Greeks and Romans were many, many centuries ahead of their time with their philosophical musings. The fact is John, everything around you, including that tiny ant scampering across the table, the air both you and the ant breath, the sun you see during the day and the moon and stars you see at night all share in common the very same, protons, neutrons and electrons that were formed during the first instant of all creation. In effect, you and that little ant are as old as the universe."

John sat in silence for a moment pondering what Neva had just said, her words burning into his consciousness. He'd never thought of creation much at all, let alone in terms that crystalized the fact that everything can be linked back to a single primary event. Through out his life, John was primarily concerned with the here and now. If asked, John would agree unequivocally with the socially correct perspective declaring that all human beings are created equal. However, after listening to Neva, John saw this humanitarian ideal to be rather trite and terribly overly simplistic. What he now realized was the fundamental truth that everything, humans, animals, plants, rocks, oceans… literally everything is made of the same basic structures that are as old as the universe itself. The new perspective Neva had just introduced to John was far more basic and fundamental than simply thinking about people as all being the same. John had received a new insight about creation providing him with a deep reverence for the connectedness of everything in a vast and ageless universe. For the first time, John saw himself as something other than John Mitchell, member of the human race. He now had a sense of himself as John Mitchell, child of the stars and part of something beyond imagination.

"So…" John began after being lost in thought for a while, "…at the atomic level, I'm related to everything and everything is related to me. If that's the case, then how are things so different? I know that I have a brain, blood, skin and other things that make me, me. However, other animals have these same things but, I don't look like a tiger or a tree. Clearly a tiger or a tree or me are different fundamentally than a rock since we are alive and the rock is not. I'm with you so far. And by the way, I am quite happy to know that I'm made from atoms that were once part of a star. Wait until my girls learn that they are made of stardust! I've known since the day that they were born that they are very, very special."

Thinking of his girls brought a smile to John's face. He could see them in his mind's eye and marveled to himself about what he and Lucy had created together….out of stardust! After a reflective moment, John's thoughts returned to the questions on his mind and he asked Neva again, "So, how do these different atoms and molecules work together to make a living creature as opposed to a dumb rock. What is special about life and how have so many different forms of life come to be? Life wasn't created at the time of the Big Bang, that's for sure."

John had just taken the next step all on his own causing Neva to give him a warm pat on the arm. "You've asked the right question," She said. "All life on earth is based on certain fundamental principles and basic parts. Of all the infinite varieties of molecules that the combination of atoms can form, some of these have taken on a very special yet, universal role, in the structure and function of living creatures. As a matter of fact, not only do you share the same atomic structure with everything in the universe, you also share a set of highly specialized structures and molecules with every other living thing that has ever oozed, slithered, swam, crawled, grew, flew or walked on earth. Just like all things are built from atoms that are all the same, all living things are built from molecules and cells that are essentially all the same. Life, while remarkably diverse, is also surprisingly redundant and it all begins with the living cell."

CHAPTER 13

Life's Basic Elements

Neva began the next phase of John's creation lesson by explaining that there are two fundamental characteristics of everything that is living. First is the ability to sustain life by converting nutrients from the environment into energy that can be used to grow, move, take in more nutrients and reproduce. The other key characteristic of all living things is the ability, or drive, to reproduce progeny. These two basic characteristics are intimately linked. Energy is necessary for growth, maintenance, repair and reproduction. And, since everything in the universe eventually wears out and dies, reproduction is essential for life to be sustained throughout the millennia.

The natural trajectory of everything in the universe is to break down over time, or become disordered. This is known as the second law of thermodynamics and is called entropy. Simply stated, both the creation and maintenance of order requires the input of energy. Without energy, order cannot be maintained and systems become more random and more disorganized. Complete randomness and total disorganization are incompatible with life. As a result for life to flourish and survive, energy and reproduction are essential.

If the production of energy and the process of reproduction are the two characteristics that distinguish living things from non-living objects, the cell and its genetic material, DNA, are the two basic units of life that differentiate the living from the inanimate. All living things including microbes, plants, and animals are composed of cells. The cell is the smallest life unit that exists. A cell is a living machine that converts nutrients from the environment into energy that the cell can use to sustain itself as well as to reproduce. While a cell is the basic unit of life, it is the DNA contained deep within each cell that is the blueprint for life. Indeed, within its DNA every cell carries a complete set of instructions for making an exact duplicate copy of itself. Microorganisms such as bacteria or amoebas, for example, each exist as single independent, self-sustaining cells. In contrast, multi-cellular organisms, from the microscopic to the majestic giant red wood or blue whale are built of many single cells that function together, often in complex and interrelated systems, to sustain the entire organism. An adult human is made up of roughly one hundred trillion cells. Since most cells are all essentially the same size, a larger organism like an elephant doesn't have larger cells than a mouse or human. Rather, an elephant, giant redwood, or blue whale would contain many more than the one hundred trillion total cells that it takes to make a human being.

While nearly all cells share basic functions of energy production and reproduction, in multi-cellular organisms some cells also become functionally and anatomically highly differentiated and specialized. Indeed, complex animals such as ants, slugs, toads, mice, men and whales may contain more than two hundred different types of cells, each type uniquely designed to carry out specialized functions that together are essential to sustain a complex, multi-cellular organism. This specialization of cells is called cellular differentiation. The amazing process of cellular differentiation and specialization begins very early in the development of an organism shortly after an egg is fertilized and begins to divide. Every multi-cellular organism and all the different types of cells it contains begin as just a single, tiny cell. Despite the cellular diversity that occurs during development of a multi-cellular organism, with very few exceptions nearly every individual cell in the body shares the same two basic fundamentals of life with all single celled organisms like amoebas and bacteria, namely the abilities to produce energy and to reproduce. This, Neva explained, was because all cells regardless of their specialized function whether they be animal or plant, contain the same basic structures used to produce energy and the same tools to utilize energy for growth, maintenance, repair and reproduction. Most importantly every individual cell from the single celled amoeba to the more than one hundred trillion cells in an adult human being contains all the instructions necessary to reproduce another identical organism.

With this basic background, John listened intently as Neva took him on a fascinating journey inside a living cell.

Cells actually come in a variety of shapes and sizes but, for the most part it's easiest to think of a cell as a bag filled with liquid. The liquid that fills cells is called cytoplasm. The word cytoplasm comes from two Greek words. The first is "kytos" which, means "empty vessel". Kyoto is now referred to as "cyto" having to do with the cell. The second word is "plasm" which comes from the Greek word "plasma" meaning something formed or molded. While a cell's cytoplasm is mostly water, within the cytoplasm floats a dizzying array of small and large molecules as well as a number of complex and highly specialized sub-structures called organelles. Just as all the parts of a clock are specialized and each one serves a unique function that is essential for the clock to function properly and keep correct time, the combination of molecules and organelles in a cell work in concert to carry out the basic functions of life. That is to say, most of what makes up a cell are generic house-keeping elements common to all cells whether they be a single celled organism like an amoeba or any one of the trillions of cells that together comprise a complete animal or plant.

Another way to think of the cell is to liken it to a complete organism. Animals, including humans, have organs such as the brain, heart, kidneys, liver, intestine, lungs, etc. All these organs must work together, each doing its own job to maintain the life of the complete being. In a similar way, a single cell contains organelles and molecules that must work together to maintain life at the cellular level. Trillions of cells in the human body working together, each one containing their own organelles and molecules working together are what constitute a living, breathing, thinking, functioning human being. Without molecules, organelles, cells, organs and organ systems, life as we know it would be impossible.

One of the truly remarkable things about life is the universality embedded within life's vast diversity. From sponges to fish, dinosaurs to birds, the giant redwood and man, all are composed of individual cells that rely on virtually identical processes to carry out the two basic functions of life, namely energy production and reproduction. The instructions for all of these basic functions are encoded in a cell's genes. Genes, as John would learn more about later, are composed of DNA. However, for the time being Neva just talked about genes as the instructions for making everything in an organism. As it turns out, the genes responsible for the basic functions of life are, for all intents and purposes, identical for all organisms across all species, great or small, intelligent or not. As John would also learn later, not only do organisms from the simple to the highly complex share many of the same basic cellular structures and functions that make life possible, but widely different organisms are built using many of the same genes. Diversity, it turns out, comes from how and when these basic genes are used and not necessarily from the presence or absence of certain special genes. For example, the same building block genes responsible for the development of fins in fish are responsible for the development of flippers in whales and legs in animals and man. Similarly, the gene responsible for the development of speech in humans is also responsible for the development of songs in finches as well as the ability of mice to learn sequences of movements required to navigate through a maze. Indeed, life and its endless diversity are possible because of the reapplication of the same key genes. Rather than repeatedly reinventing everything from scratch in each different organism, the diversity of life and its increasingly complex forms arise as a result of the reuse, control and adaptation of a number of most ancient and basic genes.

John interrupted Neva as she was talking to him about cells and DNA, "So wait just a minute Neva. I always just assumed that genes and cells in one type of animal were unique to that animal and that what makes life different and varied is the difference between genes and cells? What you seem to be saying is that not only are all different kinds atoms are made up of the same things, but that all different kinds of life are made up of essentially the same things. That's kind of humbling to think that I'm really not much different from an amoeba at some level. Just when I was beginning to feel good about myself again because I'm made of stardust, you slap me down by telling me that I'm basically the same as fungus."

"Well…" Neva smiled, "…that is basically correct. You and fungus have a lot in common and much of it is as old life on earth. You see, nature tends to not invent new things but rather it adapts and refines what already works."

"In addition to these ubiquitously common genes, cells of every living organism also contain an important collection of identical molecules and sub-structures that are required for simply staying alive and reproducing. These are sometimes referred to a house-keeping components of the cell. In contrast, highly specialized cells such as nerve cells, muscle cells, liver cells, eye cells, leaf cells, root cells, etc., have evolved to contain other molecules and sub-structures that are unique to each different cell type. These specialized molecules and sub-structures are ideally suited to enable each different cell type to carry out its own unique and specialized roll in a multi-cellular organism like an ant or man, for example."

Neva continued and explained further to John that the instructions for these universal house-keeping and specialized components are contained within the genes of every cell in an organism. Indeed, one of the most beautiful and fascinating, yet sobering aspects of modern biology is our emerging understanding of how instructions contained within the genes are selectively decoded as organisms develop, grow and mature. Selective decoding of genetic instructions enables, for example, one cell to become a liver cell while another cell becomes a skin cell, or nerve cell, or an egg. This vast and remarkable differentiation of cell types occurs despite the fact that every different cell in the organism possesses precisely the identical complete set of genes or, the instructions for life.

The process of cellular differentiation in an organism begins soon after fertilization of an egg. All complex life forms from ant to man, or flower to oak tree, begin as a single cell; the fertilized egg. Contained within the fertilized egg's DNA, its genes, are the only instructions that are ever required to build the entire adult organism. As an example, in the case of humans and most other mammals as well, by day five after the egg is fertilized the developing embryo consists of a two layer sphere comprised of about seventy to one hundred cells. This ball of cells is called the blastocyst and all mammals go through this same exact phase of development. The blastocyst is comprised of two cell types. The cell type comprising the outer layer of the blastocyst of a day five embryo will develop into the placenta, the structure that supports the transfer of oxygen and nutrients between the mother and the fetus. In contrast, the inner layer of cells ultimately develops into the fetus itself. Therefore, from the very beginning when there was just a single cell, the fertilized egg contained all the instructions to become either a placenta or a complete, individual living being. As the embryo begins to develop the cells comprising the inner layer of the blastocyst differentiate yet further to become anything from skin to heart, to eye, to brain cells. This amazing process occurs over the course of days, weeks or years depending on the species and is controlled by the selective decoding of genetic instructions that were present in the original single fertilized egg cell. Ultimately, a single fertilized egg will form an adult organism perhaps containing more than one hundred trillion cells, each cell doing its own work in its own way. Yet, despite this vast expansion and differentiation, every cell, whether plant, insect, reptile, bird or mammal, has certain basic structures and functions that are ubiquitous across all life forms on earth. Life, as it turns out, though vastly diverse in appearance and behavior, is remarkably all the same in many important ways at the cellular level.

Neva next began to talk to John about the structure of cells as she described to him how cells conduct the basic operations of life, namely producing energy and reproducing copies of themselves.

Beginning at the outermost boarder of a cell and thinking once again about a cell as a bag containing fluid in which float untold numbers of molecules and sub-structures, the bag itself serves a critical cellular function by providing a protective barrier between the inside of the cell and its external environment. Known as the cell membrane, this relatively flexible structure serves literally as the skin of the cell. Whereas animal cells are bounded by the cell membrane alone, plant cells have an extra outer layer called the cell wall which, compared to the cell membrane, is stiff or rigid. Since cells contain no skeletal structure it is the cell wall of plant cells that gives plants their solid structure. The cell wall gives celery its crunch and allows a giant redwood to support tremendous weight and stand straight.

The cell membrane is made largely of fats and proteins and is pocked with tiny pores across its entire surface. Nutrients from the external environment required for the production of energy or the building of cell structures cross the cell membrane from the outside to the cell's interior. On the other hand wastes or cellular products such as hormones, proteins, or other specialty small molecules produced by the cell pass through the membrane to leave the cell and enter the surrounding environment. Not smooth as a plastic bag, however, the cell membrane actually contains deep fissures and folds along its entire surface that run far into the interior of the cell. The reason that the cell membrane has these deep folds and crevasses is that without them the distances things would have to travel to get from the surface of a cell to its center would be otherwise prohibitively immense on the microscopic scale.

Although some nerve cells have projections that may be many feet long such as those traveling from your neck to your big toe, most cells are very small and are only visible under a microscope, ranging in size from ten to thirty micrometers. An average human hair is about eighty micrometers thick. In order to put small sizes at the molecular level of the cell into some type of reference frame for John, Neva asked him to imagine that a single cell was a giant ball 100 miles in diameter, or roughly twice the size of the state of Rhode Island. A single cell this large would have a cell membrane that was 175 feet thick. On this same scale, a simple sugar molecule comprised of 6 carbon atoms would be roughly the size of Neva's Buick. Now, if there were no folds in the cell membrane and this sugar molecule needed to get from the inside of the cell to the outside of the cell it could have to travel the equivalent of nearly 50 miles in order to reach the nearest inner edge of the cell membrane. On the other hand, if the cell membrane contained many folds reaching deep into the center of the cell, that sugar molecule may only need to travel perhaps a few hundred feet before it could cross the membrane and reach the outside environment. Fortunately, the cell is built for maximum efficiency so the transport of things like sugar molecules from the cell's insides to the outside is more like crossing a football field that going over the river and through the woods to grandma's house.

Continuing on this voyage of cellular exploration and traveling from the outside of the cell through the cell membrane and into the cell's interior, the environment inside the cell is chock full of not only specialized molecules but important sub-structures as well. These sub-structures, Neva reminded John, are called organelles. There are several different types of organelles and each type plays a critical function in supporting life. Again, reflecting the great redundancy throughout life, these different types of organelles are common in nearly all cell types whether they be human, animal or plant. Without a doubt, the three most important of these intracellular structures contained in each and every cell are the mitochondria, the ribosome and the nucleus.

Mitochondria are rod-shaped structures that serve as the power plants of the cell producing the energy required for all cellular functions. Large cells with high energy demands such as muscle or brain cells may contain several thousand mitochondria within a single cell. In contrast, some single celled organisms where energy demand is low may contain only a few mitochondria.

The production of energy is basic to all life since cells require energy for everything they do. All cells need energy to build and repair cellular structures like the cell membrane, mitochondria, proteins and DNA, as well as to reproduce (i.e. divide). But in addition, cells use energy to perform many other common activities, as well as highly specialized, work. For example, all cells use energy to pump salts across the cell membrane in order to maintain a constant and healthy intracellular environment. Highly specialized cells may use energy for such things as: generating electrical impulses in heart and brain cells, contraction of muscle, fighting infection by white blood cells, absorption of nutrients and aiding digestion by intestinal cells, secreting acid by stomach cells, and detoxifying foreign molecules by liver cells, to name but a few. Regardless of the task, however, these functions all utilize energy and the energy for these activities comes largely from mitochondria.

All energy used by cells is stored in the form of a chemical bond contained within a specific molecule called ATP which, stands for adenosine tri-phosphate. When energy is needed by the cell, special processes break a chemical bond in ATP, thereby releasing the stored energy which the cell can use to conduct its work. Almost all the ATP in a cell is made and stored within the mitochondria. As it is the primary source of cellular energy, ATP is common to all living creatures be they alive today or long extinct. That is to say, the heart muscle of a dinosaur, a human brain cell or, the simple single celled amoeba all use ATP as their primary energy source. In the case of the brain, energy from ATP may be used for the creation of a thought or the storage of a memory. In the amoeba, ATP may be used for engulfing prey or for cell division. Heart muscle cells in a T-Rex used ATP to cause the heart to beat faster as the T-Rex chased and captured some poor stegosaurus for lunch. This has not changed through the millennia and ATP has, and always will be, the primary source of energy for every living thing on earth.

"I guess I never really thought about cells needing energy, let alone where that energy came from." John said. "I know we need to eat and to breath and I assumed that food and air keep us going since without them we die." John continued. "It's kind of cool that every cell uses the same thing, adenosine tri-something..."

"Adenosine tri-phosphate." Neva helped John with the word, then said, "Just say ATP for short which is how every biologist refers to it."

"OK, ATP." John said. "What I was thinking is that ATP must come from somewhere and have something to do with air and food since as far as I know you can't order a plate of ATP in any restaurant that I know of!"

"That's right." Neva answered with a smile and a nod indicating that John was on the right track. Neva then continued to explain to John where ATP comes from.

Although cellular energy production in the form of ATP occurs chiefly in the mitochondria, the first step in making ATP actually begins in the cytoplasm with a process called glycolysis. The primary nutrients that cells use to produce energy are glucose (a 6 carbon molecule of sugar obtained through the diet) and oxygen. In reality, almost everything ingested including proteins, fats, or pasta, are broken down and ultimately converted into glucose for the purpose of energy production. Glycolysis starts the energy production process with the breaking of glucose in half resulting in the formation of a small amount of ATP and two pieces of what had been the original glucose molecule. Interestingly, this first step requires no oxygen which is why some bacteria can live without oxygen. In animals, glycolysis can provide small amounts of ATP for short bursts of energy in cases where oxygen is scarce. This might occur during a sudden burst of muscle activity before the heart rate increases and begins to deliver extra oxygen to muscles during the onset of exercise.

While small amounts of ATP come from glycolysis, the process is not very efficient and the real business of providing energy for the cell occurs in the mitochondria where oxygen is required. The creation of ATP in mitochondria begins when the two half-pieces of glucose produced during glycolysis enter the mitochondria where, in the presence of oxygen, the pieces of glucose are further broken down resulting in the formation of large amounts of ATP. Energy production within mitochondria occurs in two stages. The first stage is something called the TCA or Kreb's cycle. The second stage is call oxidative phosphorylation. Whereas glycolysis produces only two ATP molecules from one molecule of glucose, this is terribly inefficient compared to what happens in the mitochondria. In the mitochondria a whopping thirty-six ATP molecules are produced from every glucose molecule processed through the TCA cycle and oxidative phosphorylation.

In addition to ATP there are two very important byproducts, or waste, that result from mitochondrial energy production from glucose and oxygen. Both of these byproducts are intimately involved in the circle of life. These two byproducts from ATP production are carbon dioxide and water. The fact is, every molecule of oxygen you breath in and every molecule of carbon dioxide you breath out are processed entirely within the tiny mitochondria throughout all the cells in your body. Indeed, since cells use oxygen to produce energy in the mitochondria the overall process is sometimes called cellular respiration because in reality, every single cell literally "breaths in" oxygen and "exhales" carbon dioxide during the production of ATP.

Mitochondria themselves are filled with a complex, elegantly organized system of tiny membranes and proteins that function tirelessly to produce the ATP that the cell uses to sustain life. The processes of ATP production are identical for every organism that walks or crawls the earth, soars through the skies, swims to the deepest parts of the ocean, or grows in any forest or meadow. The fact is that every step you travel, every breath you take, every beat of your heart, every thought you have, every memory you store or retrieve, requires ATP. The process goes on continuously without any conscious effort on your part so as long as there is an ample supply of oxygen and glucose for your cells. Mitochondria are tireless and their production of energy for you to use requires no other input from you other than to eat and breath.

John interrupted Neva, "I had no idea there were essentially little power plants inside cells that produced energy. With all the diversity of life on earth, the system is the same?"

"Yes." Said Neva. "It's a system that works well and is universal to all living things. There is, however, one interesting twist to mitochondria and that is found exclusively in plants. In addition to containing mitochondria for energy production green plant cells uniquely contain special organelles called chloroplasts. Chloroplasts are a lot like mitochondria only in reverse! These tiny plant organelles are filled with a special protein called chlorophyll which is what gives plants their green color and also allows the plant to capture energy from sunlight. Whereas mitochondria use glucose and oxygen to produce ATP, carbon dioxide and water, the chloroplast uses sunlight, carbon dioxide and water to make glucose and oxygen. The global energy cycle of life is then complete. Plants use sunlight, carbon dioxide and water to make sugar and oxygen. Animals eat plants, or other animals, breath in oxygen and break down the sugar, exhaling carbon dioxide. Plants inhale carbon dioxide and convert it back into sugar and oxygen. Animals, and plants…one could not exist without the other. The whole ballet is played out trillions upon trillions of trillions of times every second in every tiny cell of every organism on the planet. The fact is, the carbon atoms that make the pancakes and bacon you had for breakfast this morning may have been exhaled by a dinosaur as carbon dioxide millions of years ago. This cycle has been part of creation since the very beginning of life on earth."

After thinking about what Neva had just told him, John asked Neva, "So which came first, plants or animals? It seems that one could not exist without the other?"

"That's a good question over which, there has been considerable controversy." Neva said. "The best evidence suggests that plants like we have today were not present on earth until about 100 million years ago and the first flowering plants did not appear until around the time of the dinosaurs, about 200 million years ago. However, the fossil record indicates that there were land plants at least as early as 700 million years ago and that fungi existed in the oceans more than one billion years ago. The earliest animal life probably came into existence around 580 to 630 million years ago, although it's somewhat complicated. You see, early life began as single cell organisms, much like bacteria. Some of these early life forms probably possessed properties of both animals and plants. Clearly, plants were necessary to begin producing large amounts of oxygen so simple plant life most probably appeared first setting the stage for animals millions of years later. Regardless of precisely when these events occurred, what is very clear is that all life on earth, animal and plant, began as single celled organisms. Cells, that in fact, were not unlike cells today since essentially all cells have a cell membrane, mitochondria, a nucleus, ribosomes, and most importantly, DNA."

"So the nucleus and ribosomes are the other two important structures in cells that you were going to tell me about Neva?" John asked. "Its hard for me to believe that anything could be more fascinating than these little mitochondria fellows!"

John was wrong, for as he would soon to discover, while mitochondria are the power plants of a cell, the ribosome, the nucleus and the DNA hold the very keys to a cell's power, mystery and indeed, the wonder of life itself.

CHAPTER 14
Where Life Lives

To begin with, it was important for John to understand a bit about the subject of molecular biology. Molecular biology is the study of the molecules and molecular processes that allow cells to live and reproduce. Neva explained that while a cell contains many different families of molecules including proteins, simple and complex sugars, fats, and nucleic acids, by far the two most powerful and important molecules in the cell are proteins and nucleic acids. Indeed, these two families of molecules, proteins and nucleic acids, are intimately linked to one another and together comprise the heart and soul of the genetic code upon which all life on earth is based.

Proteins, John learned, exist primarily in two major types. One type is the catalytic proteins, or enzymes. This class of proteins are responsible for conducting work or carrying out specific functions in the cell such as making ATP for energy, causing muscles to contract, or building structures like mitochondria and cell membranes. The other major type of proteins in cells are structural. In effect, the working proteins or enzymes are the hammers, saws, pliers, wheel barrows, wrenches, screwdrivers, switches, faucets and other mechanical devices that cells use to build things, change things, repair things, break things apart and transport things. Structural proteins on the other hand are, as the name implies, the scaffolding, framing, plaster, reinforcing rods, bricks and mortar of the cell. While there are some other specialized types of proteins such as antibodies and hormones, these generally do not play an important role in a single cell so for John's introduction to molecular biology, Neva concentrated on enzymes and structural proteins.

All proteins are made of long chains of small molecules called amino acids. There are twenty-one different amino acids that the cell uses to make the many tens-of-thousands of different proteins that are required for a cell to live and reproduce. The function of any particular protein is strictly dependent upon the sequence in which amino acids are linked together in the protein chain. There are literally an infinite number of possible proteins since the combination of twenty-one different amino acids is virtually limitless in potential variability. On the giant cell size scale that Neva had referred to earlier, amino acids would be about the size of an average automobile. Proteins may contain hundreds or even thousands of amino acids linked end to end like beads on a string. The link that connects one amino acid to another is a special chemical bond known as a peptide bond. Importantly, most proteins are not simple linear chains. Rather, nearly all proteins have a complex and critical three-dimensional structure. This three-dimensional protein structure is created as parts of the amino acid chain fold back on itself creating precisely consistent globular and/or looped structures. A protein that is three hundred amino acids long, for example, may actually exist in a size that is only a few dozen or so amino acids in diameter as a result of the folding and looping of the three-dimensional structure.

The three-dimensional structure of proteins is absolutely essential for proper protein function and this three-dimensional structure is determined by the sequence in which the different amino acids occur in the protein chain. Changing just one amino acid in the chain sequence to a different amino acid, or simply changing the order of the amino acids in the chain can markedly affect a protein's structure and therefore its function. The three-dimensional structure of proteins is also important because it often creates pockets in the protein that may uniquely bind, in essentially a lock and key fashion, other structures or substances such as metal atoms, small molecules or even other proteins. Hemoglobin in red blood cells is an excellent example. Hemoglobin is the protein that binds and carries oxygen and it is comprised of four identical protein chains, each one containing a pocket that holds one iron atom. The hemoglobin three-dimensional structure and the bound iron atoms are essential for hemoglobin to be able to hold and transport oxygen. Hemoglobin is able to bind an iron atom because of the protein's unique sequence of amino acids and resulting three-dimensional structure. Change just one amino acid in a critical part of the protein chain and hemoglobin would lose its ability to bind iron. As a result, a modified hemoglobin may not be able to carry oxygen. A protein's three-dimensional structure ultimately determines whether a protein is a hammer, a saw, a wrench, a brick, a wheel barrow, a reinforcing rod, or a two by four.

Whereas proteins are the tools and building materials of a cell, the other major class of important molecules in a cell are the nucleic acids. Nucleic acids hold the master blue print and detailed component instructions for the entire organism, whether that organism is a single cell, a mouse, a blue whale or an oak tree. Nucleic acids exist in two basic forms. Deoxyribonucleic acid or DNA and, ribonucleic acid or RNA. Both DNA and RNA are long strands made up of only four different building block molecules called nucleotides. In DNA, the nucleotide building blocks are abbreviated as A, T, C, and G for adenine, thymine, cytosine and guanine, respectively. In RNA, T is replaced with U, or uracil. In both DNA and RNA, the four nucleotides are linked together in long strands or chains and it is the sequence of those four nucleotides that contains the message of the genetic code.

Here Neva paused and told John she would come back to RNA later. But first, there were some important and fascinating things about DNA that John needed to understand. Without missing a beat, Neva pressed on.

DNA exists as a double strand. In order to help John visualize what she meant Neva told John to envision a tall extension ladder that had been cut down the middle of each rung. Each single strand of DNA consists of the long vertical rail and half of each rung sticking out from the rail. The half rung sticking out from each rail, is either an A, T, C or G molecule linked together by the rail. Now, for a complete double strand molecule of DNA each half of the ladder is put back together so that if the half rung sticking out the first strand were an A it would be linked to its partner half rung on the other strand containing a T. Similarly, every C on one strand would be linked to a G on the other strand. Therefore, each complete rung on the ladder is either an A-T or a C-G combination. The entire DNA may be a continuous unbroken ladder that is millions of rungs long.

Next, Neva asked John to imagine taking this long double strand of DNA by each end and twisting in opposite directions, creating a ladder that has a corkscrew appearance. This twisted double stranded structure of DNA is known as the double helix and was discovered by Drs. James Watson and Francis Crick. The discovery of the three-dimensional structure of DNA was without a doubt the most important biological discovery of the 20th century for it is within the DNA structure that the genetic blue print for life exists. For their discovery, Watson and Crick were awarded the 1953 Nobel Prize in Medicine.

John interrupted Neva saying, "I've seen pictures of what I guess was DNA in some books and magazines from time to time. I remember exactly what you are talking about Neva! The DNA looks like a giant spiral ladder that goes on and on in both directions." John had a proud grin on his face as he offered this visual image to Neva demonstrating that he had a pretty clear idea of what she was talking about.

Neva gave John a nod saying, "That's exactly right John. Cartoon pictures of DNA have become fairly common-place and they are used often as symbol of medicine or health. Anyway, its nice that most people have come to recognize a picture of DNA as something related to life since DNA is precisely where life lives within a cell."

Neva continued, talking about DNA with an almost palpable reverence as she went on to explain to John that perhaps the most important thing about the structure of DNA is that the two strands of the double helix are complimentary to one another. That is to say, all of the A's on one strand are paired only to T's on the other strand while all the C's pair only to G's on the other strand. In effect, this means that each DNA strand is mirror image of the other. For example, in a small DNA segment one strand may have a sequence of nine nucleotides ATTCAGGAC while its complimentary strand, or the other half-rungs of the ladder, would have the corresponding nine-nucleotide sequence of TAAGTCCTG.

Strand One: ATTCAGGAC
Strand Two: TAAGTCCTG

In DNA, each opposing pair of A's and T's, or C's and G's on the two strands is called a base pair. Therefore, this small DNA example segment contains nine base pairs. Amazingly, the DNA of most mammals, including man, contains a dizzying 2.5 - 3 billion base pairs depending on the species which is a lot of A-T's and C-G's!

"Whoa!" said John. "Three billion base pairs is enormous!"

"It sure is." replied Neva. "Let me tell you some even more fascinating things about DNA."

The total number of base pairs and their sequence for the entire length of DNA in an organism is referred to as an organism's genome. In the case of man the entire human genome contains approximately three billion base pairs. Within the sequence of these three billion base pairs the genome contains the entire blueprint of information necessary for the organism to develop from a single cell into a full adult as well as to conduct all the functions necessary for the organism to live and reproduce.

The Human Genome Project, completed in 2003, was a monumental effort to sequence, or decode, human DNA; that is to learn and publish the exact sequence of all three billion A's, T's, C's and G's along the entire length of human DNA. While at the time this was technically a project of daunting complexity and scope the results mean simply that scientists now know the exact sequence of A's, T's, C's and G's along the entire length of human DNA. The task was indeed huge and took a number of teams of scientists many years to complete. Since then, the genomes for a number of other species have also been decoded including, mouse, rat, dog, chimpanzee, rice and corn, as well as a large number of species of bacteria and viruses. The list of species that have had their genomes decoded is growing impressively every year thanks to improvements in sequencing technology that has made deciphering the genetic code far more rapid than when the Human Genome Project first began. Importantly, as soon as the nucleotide sequence for a species has been determined it is made publicly available to anyone with a computer and access to the internet.

A cell's DNA is housed deep within the cell in an organelle called the nucleus. A cell's nucleus is the largest of all intracellular structures, typically taking up 10 percent of the interior volume of a cell, and with good reason. On the giant cell scale Neva and John talked about earlier, the nucleus would be about fourteen miles in diameter or roughly the size of a small city. DNA itself is massive in length but fortunately, the double helix is able to coil very tightly and in a highly compact fashion that shortens the DNA molecule dramatically. This is called "supercoiling" and is essential because the three billion base pairs comprising the DNA of a single mouse or human cell would be more than six feet long if the DNA were simply a straight, uncoiled strand. To put this into perspective using Neva's giant cell illustration, if the cell was one hundred miles in diameter, although its DNA would only be as wide as two automobiles, the DNA from a single giant cell would be nearly twelve million miles long if it was completely uncoiled and laid out in a straight line from end to end!

John stopped Neva at this point and asked with a puzzled expression, "That's unbelievable! But, I always thought DNA was in the genes and the chromosomes. Also, why does DNA contain two strands? Isn't one enough?"

"Ah." said Neva. "You've hit on two very important and wonderfully elegant elements concerning the biology of DNA. Let me explain a bit more and I think you'll begin to understand."

Neva explained to John that DNA in animals and plants does not actually exist as a single continuous unbroken strand. Rather a cell's DNA is divided into structures that John correctly referred to as the chromosomes. Simple bacteria have a single chromosome where the DNA chain is linked end-to-end forming a circle. In contrast, animal and plant cells possess multiple chromosomes. In humans, there are in fact forty-six chromosomes, each existing as a pair so that there are twenty-three pairs of chromosomes in the nucleus of each human cell. Every chromosome is made of an individual segment of DNA. When it is said that the human genome contains three billion base pairs this is the number of base pairs contained in all forty-six chromosomes combined. Importantly, an exact copy of the entire genome of an organism, which in the case of humans equals forty-six chromosomes made up of three billion base pairs, is contained within each and every cell of the organism. Because this is the case, the DNA in every cell of the body contains all the genetic information necessary to form an entire new organism, identical in every detail to the original. Indeed, the chromosomes of every cell contain all the instructions any cell of that species would ever need to thrive, grow and reproduce.

In life, however, there are always exceptions and in biology, one critically important exception to this chromosome pairing rule are the egg and sperm cells. A female's unfertilized eggs and the male's sperm each contain only half the number of chromosomes that are present in the rest of the cells of the body. In humans for example, the egg and sperm cells each contain twenty-three chromosomes, not forty-six. However, when a sperm joins with an egg during fertilization, the fertilized egg cell becomes endowed with a full compliment of chromosome pairs, half coming from the sperm cell and the other half coming from the egg cell. As a result, offspring inherit exactly one-half of their DNA from the father and one-half from the mother. All children in that sense, are truly genetically modified hybrids since exactly half of a child's genome comes from each parent.

The total number of chromosome in an organism varies widely depending on the species. For instance, mice have 40 chromosomes and rats have 44 compared to human's 46. However, the size of the genome or the number of chromosomes really has little to say about either the complexity or intelligence of a species. For example, apes have 48 chromosomes, dogs and chickens have 78, rice has 24, giraffes have 62, corn has 20, wheat has 42, and shrimp have an astounding 245 chromosomes in each cell! Obviously, having more chromosomes makes a creature neither larger nor more intelligent!

Chromosomes typically look like "X's" for the most part with two DNA molecules side by side, joined somewhere near the center. One exception to this "X" chromosome shape is the single "Y" sex chromosome. In animals, including humans, males are determined by the presence of a "Y" chromosome. Every cell in a male contains an "X – Y" pair of sex chromosomes. The Y chromosome indeed looks like a "Y". In contrast, females contain a pair of X sex chromosomes. The sex of the offspring therefore is determined by the father since sperm exist in two types; one-half containing an X sex chromosome and the other half contain a Y sex chromosome. This makes sense since sperm contain only half the number of chromosomes as do other cells so when they are produced, sperm either get an X or Y sex chromosome in their half of the chromosome set of the father. On the other hand in the female, all eggs are the same and contain a single X sex chromosome from the mother. When a sperm with a Y chromosome fertilizes an egg the offspring will be male since the fertilized egg will contain an X (from the egg) and a Y (from the sperm) sex chromosome pair. Conversely, if a sperm containing the X sex chromosome combines with the egg the offspring will be female since the fertilized egg will contain only an X sex chromosome pair, one from the mother and one from the father. In reality, there is exactly a 50-50 chance that an offspring will be a boy or girl.

"I need to remind Lucy to thank me for our two girls." John chimed in. "I'd like to have a boy, but the girls are great and Lucy and I have decided to stop at two. I think Lucy especially is glad to have girls who like to dress up, go shopping and do other girly-girl stuff with her. To tell you the truth, I love it when my girls sit on my lap and give me hugs. I think girls are more into snuggling than boys are."

Neva smiled and told John he's probably a great dad, making John smile too.

Neva asked John if he would like her to continue…, "Shall I keep going John? We are now down to the heart of the matter." She said, then added, "The genes."

John nodded for Neva to continue and he listened with amazement as Neva described to him how DNA contains genes that encode for all the instructions of life.

Genes are specific, well-defined segments of DNA. Each gene contains a very precise length and sequence of A's, C's, T's and G's. Within its unique sequence of nucleotides every gene contains the instructions for making a specific protein. On average, most genes are about one thousand nucleotides long. Each chromosome contains a large number of genes and the total number of genes in the genome of different organisms varies considerably. While the exact number of human genes remains uncertain, human DNA is thought to contain between twenty and twenty-five thousand genes. Again, as with the number of chromosomes, the number of genes has no relation to a specie's size, intelligence or position in the hierarchy of life. Indeed, humans have about the same number of genes as worms, while the rice plant contains nearly thirty-seven thousand genes. What is very important, however, is that the more closely two species are related, the more likely it is that their genomes contain not only the same genes but that the sequence of A's, T's, C's and G's in their genomes is similar.

Although identification of individual genes is progressing, it remains a Herculean task to define the precise nucleotide sequence of every single gene. Learning where in the genome a gene resides and attaching a specific function to every gene is a challenge of enormous magnitude. Nevertheless, the identification and understanding of the function of each gene is an important initiative that is helping scientists learn the causes, and potential cures, for disease as well to expand our knowledge of how and why organisms grow, develop and age the way that they do.

Sequencing the genome, that is, learning the exact order of the A, T, C and G nucleotide in a specie's DNA was only the first step toward understanding the genetic code. Knowing the nucleotide sequence of the entire three billion base pairs actually reveals little about genes themselves. Each gene, being a specific and finite sequence of nucleotides buried within the genome has a precise beginning and end. However, simply knowing the sequence of nucleotides for the entire genome is like having all the letters of a great novel crammed together in sequence but without any capital letters, spaces, punctuation, paragraphs, chapters or breaks of any kind. In such a case, once the sequence of letters were available, it would then be up to the reader to figure out the individual words, sentences, paragraphs and chapters. This is exactly the situation with the genome. While scientists currently know the sequence of all three billion base pairs in the human genome this is only the first step before the real work begins of trying to unlock the genetic code and to determine which portions of the genome represent actual genes themselves. This work is critical since the genes contain the information necessary for every part of an organism to be built and to function.

"Wait a minute," John interrupted after doing some quick math in his head. "Since I talked earlier with Sam I now appreciate that three billion is a really really big number. If DNA is billions of base pairs long and you said the average gene contains about one thousand base pairs, if there are only about twenty to twenty-five thousand genes that means that most of the DNA doesn't contain any genes!"

"You're right." Said Neva. "Actually if you do the math, which is seems you have, all twenty-five thousand genes laid end to end are only about twenty-five million base pairs long give or take which, is less than one percent of the entire three billion DNA base pairs that exist in the cell's genome. It's an excellent question as to why all that extra DNA exists especially when it takes up so much space in the nucleus of the cell. Remember that the cell requires energy to build the DNA in the first place and then keep it maintained. That is a lot of wasted energy if 99 percent of the DNA that is not actual genes and serves no purpose. However, for a long time that is exactly what scientist's thought about all the extra DNA. It was believed that this extra DNA was just junk DNA and served no purpose or obvious function. Since much of a cell's DNA apparently contained no information in the form of genes the 'non-gene' parts of the genome were thought by scientists to be simply left over and useless pieces of DNA that had accumulated over hundreds of millions of years evolution. In this sense most of a cell's DNA was sort of like boxes of old photographs that everyone seems to accumulate over the years and doesn't know what to do with other than to never throw them away."

"There is one certainty where science is concerned, however, and that is that things change all the time as new information becomes available. Concepts, theories and ideas are constantly changing as new data provides further insight. As a consequence, research is now revealing that these vast spans of what had once been thought to be junk DNA actually contain a staggering array of important and fascinating genetic control mechanisms. For example, much of this non-gene DNA contains on- and off-switches. These non-gene switching and control segments of DNA are responsible for conducting a highly coordinated and intricate level of activity within the nucleus that directs when, where and why individual genes are turned on and turned off. Just like a musical score these non-gene sections of DNA control a symphony of activity that determine why one cell becomes a liver cell, another a muscle cell, where the stripes on an individual zebra will be, what the precise pattern of spots on a butterfly wings will look like, and why your nose is in the center of your face."

"And, to think that only ten years ago scientists thought that all that extra DNA was just useless junk. And do you know what else?" Neva asked John rhetorically, but before John could answer, Neva continued with a delighted grin, "It gets even more interesting. Learning about how DNA and the cell work has revealed mountains of information about how life began and has evolved on earth."

"You've got my attention." John replied, as he turned for a moment from Neva to squint at the setting sun and feel its fading warmth on his face.

"What a beautiful evening." John thought to himself.

CHAPTER 15

How Everything is Built

Neva began to tell John a little about the fascinating process of how information contained within genes is used to build an entire cell and enables the cell to live. As she had promised, DNA and the cell only became more fascinating the more John learned.

The sequence of A, G, T and C nucleotides that make up an individual gene are actually arranged in sets of three nucleotide sequences in the DNA called codons. Using the earlier example of a segment of DNA containing nine base pairs ATTCAGGAC, that small segment of DNA contains three codons, ATT, CAG and GAC. Each codon or unique sequence of three nucleotides represents the specific genetic code for one of the twenty-one different amino acids that are used to make proteins.

John immediately wondered which strand of DNA contains the codons since both strands while, complimentary, are different. In the earlier example, the complimentary strand of DNA would be TAAGTCCTG. The fact is that only one of the DNA strands contains the instructions for making a protein. This strand is referred to as "sense", because it contains the sequence of nucleotides and codons that makes "sense", as in, it can be translated into a protein. The other strand, which does not contain the instructions for proteins, is called "anti-sense" or, the opposite of sense.

Neva needed to explain next the reason for having sense and anti-sense strands of DNA before she continued with how DNA gets translated for the production of proteins. All cells contain double stranded DNA. One strand contains the instructions for all the structural and tool proteins in the organism, i.e. the sense strand. The other, the anti-sense strand, contains no information used for making proteins. The pure majestic elegance behind DNA, however, lies in the fact that this double stranded structure of DNA is what makes it possible for cells to divide and for each of the two new cells to end up with identical copies of DNA.

When a cell divides, the DNA double helix in each chromosome splits apart and each single strand is used as a template for constructing two new "daughter" strands of DNA. This process, which takes place each time a cell divides, requires: 1) energy in the form of ATP, 2) a supply of the nucleotides A, T, C and G, and 3) an enzyme called DNA polymerase. In order for DNA polymerase to create two new copies of DNA from the original DNA double helix, the polymerase begins working at one end of the DNA strand and basically unzips, or splits, the double helix into two single strands. Referring to the earlier analogy Neva had used to describe to John the structure of DNA, Neva explained that DNA polymerase splits the DNA double strand apart into the two ladder rails each with a half rung of either A, T C or G, As the DNA polymerase enzyme works its way along the DNA, unzipping the strands as it goes, it takes new A, T, C or G molecules and, using ATP for energy, DNA polymerase constructs two new double strands of DNA. As the DNA splits, each of the original parent strands of DNA serves as the pattern for production of the two new daughter strands as A's match with T's, and C's match with G's. Thus, one double stranded molecule of DNA can give rise to two new complete double helix molecules of DNA that are exact copies of the original. Each of the new DNA double helix molecules contains both a newly synthesized strand as well as a strand from the original DNA. In other words, one new DNA molecule contains the "sense" strand from the original DNA and a newly built "anti-sense" strand, whereas the other new DNA molecule contains the "anti-sense" strand from the original DNA and a newly built "sense" strand.

The process of DNA duplication occurs very rapidly and lengths of up to one thousand base pairs are duplicated every second. Because multiple DNA polymerases work simultaneously along the entire length of DNA, the full genome in a human cell, all three billion base pairs, can be duplicated within sixty minutes. DNA duplication can literally occur an infinite number of times. Each time a cell divides (i.e. reproduces) and one cell becomes two cells, two cells become four, four become eight, eight become sixteen, etc, the genome is duplicated. Indeed, after only thirty divisions more than a billion cells will exist and the single original genome from the initial cell will have been precisely duplicated over one billion times. This geometric expansion is how a fertilized egg cell can give rise to an individual organism comprised of trillions of cells with each of the trillion cells possessing an exact complete copy of the original DNA, or genome, that existed initially as a single copy in the fertilized egg.

Because cells die and are being replaced all the time, the process of DNA duplication and cell division occurs billions of trillions of times over the lifespan of an organism. Amazingly, this entire process of cellular life, death and DNA replication begins with a just single cell and its original DNA. A truly remarkable phenomenon considering that without ATP, nucleotides and DNA polymerase, duplication of the DNA could not occur and cell division would not take place.

Neva next explained to John about how the genetic code contained with the genes is translated, or read, resulting in the production of proteins within a cell. Each specific sequence of nucleotides comprising an individual gene holds the specific instructions for making a unique protein. The series of events leading to the translation of a gene in a chromosome and the ultimate production of a protein is one of the most well characterized and most fascinating processes in all of modern biology.

"You mean there is something even more remarkable than the duplication of DNA billions and billions of times?" John asked with a surprised look.

"Well I think so." Neva responded. "I agree that DNA replication is simply amazing but, to tell you the truth, for me it's the translation of the messages contained in the genes that is one of the most awe-inspiring processes in nature. Genes are spectacular that's for sure. However, unless a gene gets translated into an actual product, that is a protein, the gene is really nothing more than a set of blueprints on a shelf. In the end, blueprints are only plans of what might be and it is not until a set of blueprints is converted into something tangible that the true value of the design can be realized. I think you'll see what I mean when I explain how the genetic message in DNA gets translated and results in the assembly of a protein."

John shrugged his shoulders, smiled and said, "OK, just when you blow my mind with something I didn't know and I think nothing could be more amazing, life throws something else at me that is even cooler. It's like with my girls, every time I think life can't get better and they do something amazing, they surprise me again, over and over. I guess biology and life are a lot a like in many ways aren't they? Surprises around every corner."

Neva nodded and moved on, explaining to John how the plans that are written in the DNA get converted into the proteins that actually do the work of building and maintaining cells.

The first step in the process of protein construction is the production of the other nucleotide molecule known as ribonucleic acid or, RNA. This is accomplished through translation of the nucleotide sequence in DNA. Unlike DNA, however, RNA is a single stranded molecule. John recalled that Neva had said that in RNA, the "T's" from the DNA are replaced with "U's" or uracil. Just like DNA duplication, however, the production of RNA requires ATP as the source of energy, a supply of nucleotides (A, U, C and G) for making the RNA, and another enzyme called RNA polymerase which does the actual work. The process begins inside the nucleus when a specific segment of DNA receives a signal indicating that it is time to make a certain protein. Once the signal is received, the RNA polymerase finds the exact site on the chromosome where the specific gene resides, unzips the double stranded DNA at the site of the gene to be translated, and begins to construct the RNA. As the strand of RNA is built and peals away from the DNA, the DNA itself goes back (re-zips) to its original double stranded form. The newly made single strand of RNA is called tRNA or, transfer RNA, since this piece of RNA must next travel to the outside of the nucleus into the intracellular space where proteins are made. The entire process of un-zipping the specific gene, making tRNA and moving the tRNA out of the nucleus and out into the cytoplasm requires only minutes to complete after a signal is received and a specific gene is "turned on". Turning a gene on which involves the processes of translation of a gene into RNA and then ultimately conversion of this genetic message into a protein is known as gene expression.

Scientists have learned a great deal about how gene expression is controlled. It is now recognized that the process of gene expression is directly responsible during embryological development for causing some cells to become liver cells, while others become skin cells, and still others become heart cells, etc. Even more importantly, however, the control of gene expression during embryonic development determines the distinctive and differentiating physical traits of a species as well as the specific distinguishing characteristics that are unique to each individual. Examples of general characteristics common to a species include such things as whether an organism has legs or fins, hands or hooves, stripes or spots. In addition, and even more remarkable however, is the fact that precise control of DNA expression is also responsible for unique characteristics that differentiate one individual from another within a species. Unique characteristics include such things like the specific pattern of stripes on each individual zebra, the precise size, shape and location of spots on an individual butterfly's wings, the size and shape of your nose, the thickness of your eyebrows, and the birthmark on your ankle.

In order to accomplish such remarkable and specific actions, DNA is endowed with a complex network of switches and control points that recognize specific signals arising from inside as well as outside the cell. DNA switches are actually proteins that rest on specific areas of DNA and function to convert internal and external signals into precise patterns of gene expression by literally turning some genes on and turning other genes off at highly specific times and for specific periods of time during development. The orchestration of signaling within DNA is ultimately responsible for creating the more than two hundred different cell types that comprise complex multi-cellular organisms as well as for uniquely differentiating one individual from another. These signals "know" which strand of DNA is sense and which is anti-sense and direct the RNA polymerase to only copy the sense component of genes. The many types of cells in an organism can differ remarkably in terms of size, shape, life span, and function. However, in an individual organism everyone of it's cells contains the exact same DNA as every other cell. Cells only become different, a process called cellular differentiation, because different switches and DNA control mechanisms have orchestrated the expression of specific and select genes in each individual cell. Therefore, while all zebras have stripes, the distinctive number, position, width, and length of each individual zebra's stripes are controlled by the expression of genes during development that make each zebra's stripes unique to the individual.

The process of cellular and individual differentiation resulting from the dynamic control of gene expression is a relatively new area of biology known as evolutionary developmental biology or, "evo devo" for short. Evo devo is uncovering the wonderfully elegant and beautiful control mechanisms buried within DNA and responsible for creating an endless variety of organisms and species. Although many organism share similar if not identical sets of genes, the control of gene expression decorates all of nature with a boundless variety of colors, shapes, sizes, patterns, functions and behaviors. The control of the expression of similar genes to create the seemingly endless variety of life found in nature is truly one of miracles that molecular biology has uncovered through exploration of the genome. Indeed, the more that scientists learn about the control of genes, the more fascinating this branch of science becomes. Neva promised John she would come back to the control of gene expression later when they talked about the process of evolution. For now, however, Neva wanted to finish explaining to John how the specific message translated from a gene into RNA gets converted into a protein, for it is the protein that is ultimately responsible for carrying out the function of the gene itself.

Once the tRNA is synthesized from the DNA template and exits the nucleus, the RNA next combines with an organelle in the cytoplasm called the ribosome. The ribosome is literally a molecular machine comprised of a collection of several structural proteins and enzymes. The process of protein synthesis by the ribosome begins when RNA, carrying the message from the gene, enters one end of the ribosome. Then, using ATP for energy and amino acids as the building blocks, the ribosome literally manufactures a protein molecule by linking amino acids together in the correct order as the codon sequence of the RNA is read. Amino acids and RNA enter one end of the ribosome and the newly synthesized protein molecule is cranked out of the other end of the ribosome. As this process of protein synthesis occurs, the chain of amino acids folds in precisely the correct fashion as it exits the ribosome. Once complete, the newly constructed protein is transported to its place of function within the cell.

Not surprisingly, because proteins in cells do physical work, that is, they are either enzymes (tools) or they are serve a structural function, all proteins eventually wear out. For this reason there are wonderfully efficient mechanisms in cells for identifying, breaking down and recycling old proteins, replacing them with new ones. Living cells are in fact, the ultimate model for ecologically efficient recycling. As a result, the entire cell and all its internal parts including the nucleus, mitochondria, cell membrane, ribosomes, and other components are constantly being built, broken down and rebuilt with recycled parts. The entire process is controlled through the expression of genes, orchestrated by various switches and control mechanisms within the nucleus.

John interrupted Neva, "It seems to me that here is another chicken and egg issue." John said while contemplatively scratching his chin. "I know that that life on earth may simply have begun over millions of years as molecules like amino acids and nucleotides formed in the oceans. But, if cells require energy in the form of ATP, and the production of DNA and proteins requires energy, nucleotides, amino acids and enzymes, where did the original DNA and protein come from in the first place when life began? It seems that not a single protein could have been made without first having a minimum…" John paused as he tried to remember some of the details of what Neva had been talking about. Then, touching each of his fingers to his thumb in succession as he spoke, John continued, "...a gene, RNA polymerase, a mitochondria and a ribosome. So where did the protein for the polymerase, mitochondria and ribosome come from in an ocean were some amino acids may have been sloshing around?"

"Ahhhhh!" Exclaimed Neva with a twinkle in her eye. "You've hit on perhaps the most significant issue that should frame any debate between the origin of life and the process of evolution. You see, in the case of existing organisms, that is existing life, the original DNA, protein, ribosomes and mitochondria for getting things started already existed in the egg. The fertilized egg already contains all these structures. After fertilization, the control mechanisms are kicked off and new DNA, protein and more mitochondria can be made. That's fairly simple to grasp since the basic elements of life already preexist in the fertilized egg. But, go back ten billion years when the earth was first formed and no life preexisted. If life supposedly was to have just begun in some primordial soup, where did these original tools, raw materials and energy come from? The fact is that no one really knows. It doesn't take much knowledge of cellular function, DNA replication and protein synthesis to understand that that it would be remarkable indeed for all this to have just happened by random chance. Certainly, its a heroic leap of faith in statistical probability to conclude that the random collision of molecules over billions of years occurred in just the right manner and that, as a result, life just began."

"Hmmmm." Mumbled John, nodding in agreement. "That is quite a stretch when you think about it."

"Yes it is." Said Neva smiling. "It's a debate that often gets confused with evolution. It's one thing to envision life adapting over time to changes in the environment for example. Over eons of time as living creatures encounter an ever-changing planet, life could indeed gradually morph into different forms and structures which is the process of evolution. But, its quite something else all together to consider that life itself just sort of happened purely by random chance. Now that you have a basic understanding of how the fundamental inner workings of a cell function, we can talk about the core issues of the origin of life and evolution in a context that hopefully will make some sense. That is, if you really want to keep going." Neva stopped and gave her shoulders a little shrug in a non-verbal suggestion to John that he was welcome to call it quits with his biology lesson and change the subject if he wished.

"Just fascinating." John said with a shake of his head while seeming not to notice Neva's subtle offer. "You are really good at explaining this stuff. If I'd had had someone like you teaching me in school things might have been different. You never know, I might have become a molecular biologist if you had been my biology teacher!"

Neva laughed and patted John's arm affectionately. "Better late than never John. Perhaps I'll create a convert. New science recruits who grasp some of the basics and who are willing to look at creation in a more enlightened fashion are always a welcome to learn from all that nature and creation have to offer. You see, everything around us has always been here to explore. It's just that some choose to take it for granted or refuse to try and understand. The answers are here." Said Neva, again holding up her hands and gesturing to everything around her. "It only requires you to look, explore, and to think." Realizing that John wanted to know more, Neva took John on a journey back to the dawn of time when the earth was a boiling cauldron of molten rock upon which no life in any form could have existed.

CHAPTER 16
Back in Time

The crowd of spectators continued to grow. Once again he slipped into a state of semi-consciousness and time began to accelerate. Faster and faster in his mind's eye; that link to one's inner self, one's imagination, one's dreams. The image of self that everyone carries in their head. Inside himself, somewhere deep within his mind he began a journey through time…backwards.

The day's events began to unfold like a movie running in reverse. Driving south, a beautiful day, the road clear. Meetings throughout the morning. Conversations. He could see faces of colleagues and hear their voices as time rolled backwards. Slowly accelerating, prior weeks and then months passed. The birth of his children, his wedding, college, third grade.

Faster and faster time rewound. Suddenly, he was sitting atop a high mountain, higher than anything else around him. It was freezing cold. He squinted painfully, blinded by an endless expanse of ice and snow that reflected light from a sun suspended in a cloudless sky. The earth's landscape spread out before him in all directions as far as his mind's eye could see. From this lofty perch he watched as huge mountain ranges self-assembled from sand and rubble, then at their acme those mountain ranges were sucked back into the depths of the earth from where they were born. Massive canyons, miles deep and hundreds of miles long filled as mighty rivers disgorged the rock and sand that the rushing water had eroded away over eons of time. Immense glaciers covering more than half of the planet retreated to the north and south poles, then expanded again in all directions as masses of snow and ice miles deep tightened their frozen grip on the planet once again. Giant lakes emptied, then filled with rock, gravel and sand. Mountains that had been ground to rubble were reborn as glaciers repeated their cyclic dance over the face of the planet. Great oceans flooded the landscape. Then even more rapidly than they had appeared, the seas receded. Layer upon layer of sediment were sucked back into vast oceans as the water expanded and contracted across an ever-changing landscape. Volcanoes, too many to number, decorated the earth as far as the eye could see. The fiery mountains inhaled deeply as ash, rocky rubble and lava thousands of feet deep covering millions of square miles of landscape where sucked back into the earth from where it originated. Over and over again the cycle was repeated as volcanoes inhaled then belched the earth's molten interior miles into the sky and far over the land, enshrouding entire land masses under thousands of feet of volcanic waste killing every living thing. Continents, at first familiar in shape became completely foreign in appearance as land masses drifted endlessly over the surface of the earth like great vessels upon an unseen ocean. Massive bodies of land assembled into a single huge continent and was then ripped apart as horrific volcanic explosions cracked the earth's crust like an egg shell. Continents crashed violently into one another then parted once again leaving huge scars on their landscapes in the form of high mountains, deep valleys and endless plateaus. Great storms, hurricane force winds, lashing rain, snow, ice, and lightening molded the earth all around him as he traveled back in time. Millions, then hundreds of millions, then billions of years passed before his eyes in what seemed to be only a few moments in time.

    Massive jets of gas and molten rock that had erupted violently from the surface of the earth were sucked back into a forbidding landscape. The mountain that had served as his perch was suddenly gone and he found himself suspended, floating weightlessly above a boiling cauldron of molten rock. Burning, churning fury. Searing heat rose from everywhere, endlessly. Horrific explosions shook the earth as the planet inhaled plumes of molten fire, rock and ash that had been hurled hundreds of miles into space. Gasses from deep within a young planet violently expanded and exploded from the surface with tremendous force. Meteors, asteroids and debris of all shapes and sizes that had rained down from space unceasingly over eons of time and in uncountable numbers were violently jettisoned back into space as time continued to accelerate backwards. Colossal projectiles from space that had been absorbed into the molten planet sending gigantic crowns of liquid rock far out into space were hurled back into the cosmos. Some of the debris from these impacts returned to the infant planet while other mammoth pieces broke free from earth's gravity and hurtled off into the void. He watched stars, millions of times larger than the sun, form from giant gas clouds that expanded and contracted into burning orbs only to explode, instantly sending remnants far out into space before imploding again as stars died and were reborn. Galaxies by the billions collided violently then raced apart. The entire universe itself was a living breathing organism. Each cosmic breath creating and destroying countless numbers of galaxies, stars and planetary systems.

As he floated above the now strangely peaceful vista that continued to shrink around him, he became dimly aware of a single point of light. At first faint and distant, he was drawn closer and closer to the light which became stronger and painful to look upon. Shielding his eyes with his hand, he tried unsuccessfully to turn away from the blinding beacon. Try as he may, the light became impossible to avoid since it was all around him and in fact, the light seemed to be radiating from him in all directions in an endless spiral resembling the arms of a galaxy or the chambers of a conical seashell. He became confused and disoriented. While the light was bright beyond anything that he had ever experienced and was painful to look upon, the light also beckoned him with its hypnotic quality. As the hypnosis wrapped seductive fingers around his mind's eye the light became even more intense. Strangely, the more intense the light grew, the more comfortable it became for him to look at. No longer having to squint, he removed his hand from in front of his eyes. The light drew him to it, into it, warming him. As he slowly floated toward and into the light, his mind's eye began to change. No longer was his sense of self in his head. The constant drone of self talk that had consumed his consciousness throughout his life for as long as he could remember suddenly became deafening silence. There was no sound. There was no longer light. There was no longer space. There was only complete nothingness, tranquil, peaceful and empty. For the first time since he had become aware of himself he was no longer listening to the voice in head because he had become that voice. He felt fantastic.

CHAPTER 17

The Greatest Leap of All

"Fourteen billion years ago…" Neva continued.

"Excuse me? I'm sorry, what did you say?" John said, then adding. "Sorry my mind drifted for just a minute there. I was looking at the sunset and kind of got lost for a moment."

"Oh no problem John." Neva said. "That is one of the more beautiful sunsets I've seen in quite a while. It truly is a beauty isn't it. Anyway, I was just saying that fourteen billion years ago is when everything is supposed to have gotten started." She repeated. "According to the Big Bang Theory, nearly fourteen billions years ago a single point of pure energy exploded giving birth to the universe as we know it. The release of such a massive amount of energy caused things to happen relatively fast during the first million years or so. The universe expanded rapidly and cooled as a result. The first atoms, hydrogen and helium were formed then began to collapse, forming stars by the billions as the first galaxies took shape. These young stars eventually died, exploding and forming new stars around which planetary matter accumulated and solar systems were born after the first few billions of years. It is thought that earth first formed around ten billion years ago or about five billion years after the Big Bang."

"Earth is not very old then, that is at least in billion-year terms?" John asked.

"That's right." Neva said. "Billions of years is nearly an incomprehensibly long period of time to fathom from a human perspective. There is no reason to believe that the universe will not be here for billions of years yet to come. In that sense, at ten billion years old, earth is a mere child in the cosmos!"

Neva continued, "When the planet earth was formed about ten billion years ago it was a molten ball, constantly bombarded by remnants of a stellar explosion as the solar system containing the earth and other planets around the sun began to take form. Over hundreds of millions of years things eventually began to quiet down allowing earth's crust to cool. The essential elements of life, water and oxygen appeared. Then, sometime nearly four billion years ago the earth became hospitable. In other words, conditions became ripe for the earth to support life."

"Ahhhh." John said. "The primordial soup! But, where did the water and oxygen come from in the first place?"

"There are a number of theories about this, bringing up yet another chicken and egg conundrum." Neva responded. Her deep, piercing dark eyes twinkled. She clearly loved the conversation she and John were having. She taught. John thought.

"Oxygen might have either come from plants, converting carbon dioxide into oxygen, or from the breakdown of water molecules into oxygen and hydrogen by ultraviolet radiation. There is ample evidence to suggest that earth's early atmosphere was loaded at one time with carbon dioxide. This original carbon dioxide came from massive, global and violent volcanic activity as the earth cooled and formed a thin outer crust. Early plant life could have been responsible for converting this carbon dioxide into oxygen, however, for plants to develop and grow they would need water and oxygen just like any other living thing. Therefore, to get around this complication some scientists instead favor the idea that as earth was bathed in ultraviolet radiation from the sun, water vapor in the early atmosphere was broken down by this radiation into hydrogen and oxygen. Obviously the question becomes, where did the water come from?"

"Water is also known to be produced by volcanic activity along with carbon dioxide and a variety of other gasses such as ammonia and methane. Certainly, volcanoes ruled the earth in the beginning millennia of the planet's existence. As a result, volcanoes are likely responsible for creation of much of the earth's early atmosphere. In addition, volcanoes helped to keep the earth warm four billion years ago which is probably a good thing since at that time the sun was not nearly as hot as it is today. Imagine a planet literally covered with volcanoes and its not hard to see that they could spew out unimaginable amounts of water vapor and carbon dioxide into the atmosphere."

"Yet another theory about the origin of water is that water arrived on earth from space. Giant asteroids and meteors made of snow and ice bombarded earth relentlessly during its formation. Because earth was so hot the water in these cosmic ice bombs would have been instantly vaporized as they smashed into the earth, releasing huge amounts of water vapor. Of course, the question then becomes where did the water that comprised the ice projectiles come from?"

"The point is..." Neva concluded, "...that nobody really knows for sure. These different theories all have their strong and weak points. Regardless, what is quite clear from the geographical record contained in earth's rocks is that somewhere around four billion years ago the earth finally had an atmosphere comprised of water and oxygen in sufficient quantities to support early life."

Gazing thoughtfully at the sunset that had seemed to grow more glorious with each passing moment as Neva had been speaking, John asked, "So by whatever means water and oxygen came to be, life supposedly started in the oceans, lakes and puddles billions of years ago?"

"Supposedly." Neva replied with a curious knowing look as if she was intentionally withholding information so as to let the drama build and keep the mystery growing. "You see, there are two great questions that puzzle scientists today. Neither can ever be definitively answered so there will always remain an element of mystery surrounding the origin of life on earth. There are a great number of well known and understood theories in science such as the theory of gravitational force, the theory of relatively, the theory of thermodynamics, and the theory of evolution, to name just a few. Each of these theories has, over time, been tested and retested and the data consistently suggest that these theories are correct. Scientists continually challenge their theories with predictions and conduct experiments or record observations of the physical world in order to examine how well theories hold up. The more this is done, the more we can become certain that these theories correctly explain how things work whether that be gravity, time and distance, or the evolution of life on earth. Until data is found to the contrary that definitively argues against these theories, they are considered to be correct."

"When it comes to the Big Bang there is really no way to test how the universe was created. Similarly, there really is no way to confirm how the origin of life on earth occurred."

Neva went on, "It is impossible to know for certain where the nucleus of pure energy came from that caused the supposed Big Bang. In fact, there is no way to know if this is really what happened at all. Certainly, all the available data suggests that the Big Bang is the most likely explanation for the origin of the universe. Scientists may speculate endlessly about how the universe came into being using both the known laws of physics as well as knowledge learned to date about how stars are born and die. However, if the Big Bang does seem to be how things started, at the end of the day no one can know with any certainty about the origin of the single source of energy that caused the Big Bang."

"Was the energy made, or had it always just been there? If so, where did it come from in the first place? If it was made, who or what made it and why? How long before the Big Bang did the energy exist? If the energy was made, who or what made the maker? Then, who, or what, made the maker's maker…and on and on it goes without end. Arguments about the origin of the universe, just like the origin of life, naturally progresses rather quickly into a philosophical discussion rather than a scientific discourse based upon facts and data. There is really no way to avoid it."

Neva's face beamed as she talked in a voice that could only be described as hypnotizing. Neva's voice was an auditory hallucinogenic drug to John, quiet, soothing, low and sultry. She sounded like the ubiquitous robotic female voice in transportation hubs of the future that calmly and confidently directed busy travelers when to board their transport vehicle or where to proceed to catch their next flight to some distant world. John detected a slight accent as well. South African, or perhaps Australian or New Zealand? John couldn't be quite sure. What he was sure of, however, was that he could listen to Neva all day. Just the sound of her voice made him feel warm and comfortable.

As Neva continued, John's mind replayed what she had said, *"Billions of years, pure energy, a giant explosion, stars, planets. Galaxies far too numerous to count. Distances that are impossible to comprehend, unending vastness?"* John thought to himself….then shaking his head he said to Neva, "I've never thought of things like this before. I can't get my head around the incomprehensible expanse of it all."

"Fortunately you don't have to." Neva replied. "Nobody has come up with a way to make it simple. Sometimes its enough just to know that it can't be known and accept that in the beginning, creation happened and the laws of physics, chemistry and biology were set into motion."

"I guess you are right. So, you said there were two unanswered questions?" John continued, "…and I bet the second is how life got started in the first place?"

"Right you are." said Neva as her eyebrows arched and her eyes sparkled, reflecting the pinks, blues and grays of the stunning sunset.

"About 4 billion years ago the environment on earth was finally ready to support life. A careful look at the fossil record tells us that around 3 to 3.75 billion years ago, single cell organisms first appeared on the planet earth. This would make some sense since a single cell is certainly the simplest form of life. If indeed life began somewhere in some form, it would logically be a single cell to start with. However, for the longest time, roughly the next 3 billion years, this was all there was...that is, just single celled organisms. It took 3 billion years before more complex multi-cellular organisms began to appear in the oceans. What caused single cells to become co-dependent upon one another and diversify into more complex multi-cellular organisms is completely unknown. It's as if, for whatever reason, single cells decided that two cells must be better than one and, three are better than two, etcetera, etcetera!"

"Going back to the very origin of life itself, however, there are basically two schools of thought." Neva posited holding up one finger. "One viewpoint argues that over hundreds of millions years in the primordial soup, spontaneous and purely random chemical interactions resulted in the formation of the basic building blocks of life, namely, amino acids and nucleic acids. Over eons of time these molecules combined further to form simple peptides (i.e. protein fragments) and small segments of DNA or RNA. Given the eons of time involved it can be argued that it was purely a matter of statistical probability before these molecules gave rise to some form of primitive life from which all other life could have evolved. There is even some modest evidence for this idea. Scientists in the laboratory have created some basic simple molecules like amino acids by using simulated lightening along with a primitive gaseous atmosphere and a concocted man-made primordial soup. How closely these experimental conditions resemble the starting conditions on earth is only an educated guess, however. On the other hand, while there is fairly good evidence that such conditions could have led to the formation of amino acids, there is considerable controversy over whether or not these same conditions might have led to the creation of the nucleic acids necessary for making working molecules of DNA and RNA. As a result, there is a great deal of uncertainty and ambiguity about how life could have originated from the random sloshing about of simple molecules in primitive oceans over billions of years."

Now holding up two fingers for emphasis, Neva continued. "The second idea regarding the origin of life on earth invokes the concept of extraterrestrial life arriving on the planet. These theories suggest that perhaps primitive single cell life forms arrived on earth by hitchhiking on a meteor or asteroid. Alternatively, perhaps the building blocks of life, that is, amino acids, nucleic acids or their precursors came to earth aboard a meteor crashing into a pond. Of course, this is another classic chicken and egg conundrum since if life first arrived on earth from somewhere else, how and where did it start originally? Proponents of the first idea, that life just sort of got started over millions of years of random chemical events argue that is precisely how life would originate anywhere. And so goes the endless argument.

"Isn't there a third idea?" John interrupted. "What if all life was simply created exactly as it exists today? It doesn't seem that there is an easy, or really, any answer for how life got started." Mused John while looking at the sunset. "It's all such a long time ago and life is so diverse and complex. I just don't see how there can be any reasonable explanation for how life began and then evolved into so many different forms. Unless everything was just made the way it is today, how could life start out with just a few simple single cells and then ultimately result in the menagerie we have in every zoo around the world?"

Neva smiled. "John, there is an explanation for at least part of your question. While it is impossible to know for certain how life itself began, what is very clear is how life has evolved or changed over time. On this subject, a great deal is actually known. Evolution may be one of the most important scientific discoveries ever because with this knowledge, it is possible to literally peer into the eye of creation itself. When you understand something about the theory of evolution you simply cannot help but stand in awe of life's unending ability to change, adapt and diversify. You see, life itself is a powerful creative force that is perfectly tuned to diversify and continuously experiment with form and function. Several times throughout earth's history, the tree of life has been pruned back to nearly the bare trunk by climactic change or meteor impacts that have devastated the entire planet. More than once, mass extinctions have occurred as millions upon millions of species were wiped from the face of the earth never to appear again. However, after each catastrophe, life showed its marvelous resiliency and uncompromising sustainability. Refusing to let go of its grip on earth each time a global cataclysm destroyed nearly everything, life brought forth new and even more remarkable species than those that had disappeared. How humbling to be part of a single tree of life on whose branches the process of creation continues through millennia without end. Rather than stimulating controversy, a simple understanding of how evolution works should engender a profound reverence and sense of wonderment for the improbable gift of life that is shared by every creature on the planet."

CHAPTER 18
One Tree, Many Branches

Neva wanted to make sure that John understood the difference between the ORIGIN of life and the EVOLUTION of life. Evolution, as first hypothesized by Darwin in 1859 describes the branching, differentiation, modification and adaptation of life over time. While Darwin correctly recognized through his studies that over long periods of time species diversify into new forms, he did not understand or know about the mechanisms by which this process occurred. It was not until many years after Darwin published his landmark book "On the Origin of Species", that DNA was discovered. It is now recognized that the modification or mutation of DNA is the means by which evolution occurs. DNA must exist in order for evolution to take place because it is through DNA that all traits become manifested in an organism. Physical and functional traits become modified over time and those changes are then passed from generation to generation in the form of small modifications at the molecular level of DNA, leading to observable changes in biological structure and function. Continuously adapting and changing over thousands, millions and billions of years, life evolves.

In contrast, the question of the origin of life deals with the issue of where DNA came from in the first place. In other words, for evolution to occur, DNA must first exist. For DNA to exist, either it was formed randomly over millions of years or something else happened.

"Wait a minute." Interrupted John. "I can see that for evolution to occur DNA must be present. I can even understand that by pure random chance it might be possible over millions and billions of years for a few molecules to bump into one another in just the right way to form an amino acid or a nucleic acid. I can even buy that this might happen countless of billions of times over hundreds of millions of years. However, I thought that for life to exist that a cell needed to have the ability to produce energy and reproduce. Didn't you just tell me that the production of ATP and the duplication of DNA requires specialized structures, enzymes and proteins? It seems quite a jump to think that all these pieces of the puzzle randomly came together and just started working even in a simplistically crude fashion by pure random chance over hundreds of millions of years! I just don't see how its possible to make the leap from a few amino acids being formed in an ocean to a single cell that can generate and use energy to make a duplicate copy of itself. That seems like a real stretch of the imagination."

"Well..." Neva responded with a knowing smile. "...yes, you have put your finger on perhaps the greatest mystery in all of science. Even the most critically thinking and speculative scientists have no answer for the fundamental question of how it is possible to make the leap from a sloshing soup of molecules to a single living, energy producing, self-reproducing cell. While nearly 100 percent of scientists today accept the theory of evolution and the role that DNA plays in initiating and passing along inherited changes in biological structure and function through generations, there is next to no agreement or even any reasonable way to test how life itself began. We know for certain that life began billions of years ago. On that there is no doubt. However, while there are theories galore on how that actually occurred, it is simply beyond our ability to know for certain how exactly life began. All of the plausible theories on the origin of life itself rely on assumptions of both perfect conditions coupled with statistically complete randomness over eons of time. The odds of any of these things actually happening are too large to comprehend. With absolutely no way to prove definitively how life actually began, I find it is easiest and most comforting to simply accept that life on earth began several billion years ago and, most importantly, life has been adapting, changing, and evolving ever since."

Neva continued. "Did life begin spontaneously, or was life intentionally created as a single, simple cell? Spontaneous development, or intentional creation...each comes down to key assumptions and neither can explain the beginning of life with any degree of certainty. What is very clear, however, is that once the first kernels of life came into existence as rudimentary cells possessing all the components necessary to produce energy and duplicate DNA, the stage was set for evolution to occur. Once life began, the tree of life was planted and it has grown with increasing complexity and endless diversification that will continue for as long as life has a foothold on earth."

John nodded indicating he understood and was with Neva so far.

"In fact, it is essential to think of evolution as a tree or branching bush and not as a ladder. Unfortunately and quite mistakenly, a ladder is how most people think of evolution today. Evolution is not a continuous straight-line march, or climb up the rungs of a ladder toward increasing complexity, improvement and intelligence. The classic example of this erroneous belief is the famous monkey to man mural that has been adapted into many forms and appears in some modern science education books. I'm sure you've seen this series of images. It begins with the figure of an ape-like creature which transitions in discrete steps to ape-man-like figures, each one becoming increasingly upright, taller and more human-like with every step. Sometimes one of the images includes a caveman that may be carrying a club or spear. Finally, the last figure is often depicted as a modern man in a business suit carrying a briefcase, looking intelligent, walking perfectly upright and proud. This simply could not be further from the truth."

"The fact is that humans branched off from monkeys nearly five million years ago. Monkeys continued to evolve along their branch while human evolution continued along its own branched path. Over five millions years of evolution, human life has experimented and nurtured those characteristics that make humans uniquely human whereas monkeys have evolved along a separate branch reflecting the characteristics that make them uniquely non-human primates. The process of evolution leads to many dead ends as new organisms or species sprout from a main branch only to end in extinction; that is, failed experiments. In truth, more than 99 percent of all species of animals and plants that have ever existed on the planet earth are extinct today including a number of human species. In this sense, evolution is a process of continuous branching, trial and error, adaptation, change, life, death, proliferation and extinction."

"Humans?" John said, his head popping up as he turned toward Neva, looking squarely into her dark eyes. "Are you telling me there are humans that are extinct? I thought that we were the crowning achievement in creation. That we were special? Even if we did evolve from monkeys, I thought we just continued to advance and left chimpanzees in the dust. But you said the human race evolved from trial and error?"

"Well John…" Neva replied. "…there are those who believe in a creator, a God, that made man special compared to all other forms of life on earth. Some even believe that man has been made in God's own image. Well, if a God did indeed create man in His own image and made man special, that God threw away a lot of prototypes in the process. The fossil record is absolutely clear on this. In fact, there are more than a dozen different species of humans that have been identified. In technical terms these are called 'hominids', which means modern humans and their ancestors. Although all these different extinct human ancestral species as well as modern day humans share 99 percent of their genes with chimpanzees, hominids either extinct or alive today are distinctly different from monkeys. Different hominid species can be identified through fossils dating back nearly three million years. Many of the human, or hominid species that lived five hundred thousand years ago were known to use primitive tools. They also practiced rituals that we today might call religious, such as burial of their dead and offering gifts to the gods. So in fact, many of these extinct human species behaved in many ways like modern man."

"More importantly, however, several of these now extinct species of human actually lived on earth at the same time. The best example of this are the Neanderthals who lived from about 250,000 BC to as recently as 30,000 BC. *Homo sapiens*, or modern man, have been on earth from about 200,000 BC to the present. Neanderthals lived mostly in Europe, Asia and parts of the Middle East. *Homo sapiens*, on the other hand came out of Africa and migrated throughout Europe, Asia, Australia and ultimately North and South America. The last Neanderthals died off around 30,000 years ago, while *Homo sapiens* flourished and have continued to evolve into modern day humans."

John's mouth was agape. He stared at Neva in disbelief as if he'd just heard something that shook his mind to its very foundation.

CHAPTER 19

One is a Lonely Number

"I had no idea." Said John, after a taking a moment to partially digest what he had just learned from Neva. "Are you sure you are not making this up just to mess with me?" John added in mock jest with a shake of his finger in the air. John knew with absolute certainty from Neva's vast knowledge, gentle demeanor and quiet soothing voice that she was completely incapable of telling even the smallest of white lies.

John continued, "So you are saying that as man evolved, or branched away from monkeys on the evolutionary tree some five million years ago several early species of man didn't make it! Some humans simply died off while others continued to adapt and change? I had always assumed that Neanderthals were just the cave-man ancestors of humans. I had no idea they were actually some other species that might have lived side by side with our real ancestors! Did we ever bump into one another? What happened to the Neanderthals? Could I be related to one! I'm sure Lucy would not be surprised to find out I have some cave-man DNA." John added with a laugh.

"It's not exactly clear what happened to the Neanderthals." Said Neva. "The Neanderthal population is believed to have been relatively small compared to *Homo sapiens*, and probably never reached more than ten to twelve thousand individuals spread across Europe at any one time. There is not good agreement on why the Neanderthals disappeared but theories suggest competition from *Homo sapiens*, lack of food, cold, disease, an inability to communicate as effectively as *Homo sapiens*, or a combination of these led to Neanderthal's ultimate extinction. Whatever the cause, Neanderthals, as well as several other archaic species of primitive humans became extinct and left *Homo sapiens* alone as the sole humanoid species some thirty thousand years ago. What is very clear, however, is that DNA analysis has shown definitively that while Neanderthal DNA is distinctly different from both human and chimpanzee DNA, recent genetic analysis suggests that humans and Neanderthals may have interbred. So, in essence you very likely have some Neanderthal genes mixed in with your *Homo sapien* DNA. Apparently, this should come as no surprise to your wife Lucy!"

Neva smiled and stopped.

"So why did I never learn any of this in school!" John exclaimed with a sarcastic air. "I've just always thought that first came monkey's, then cave men, and finally us. And now I learn that different kinds of humans and pre-humans roamed the earth at the same time. All of the other died off and you and I are what is left…at least for the moment. And, DNA from Neanderthals has been analyzed! How'd they manage that?"

Not really expecting a response since he'd just parroted what Neva had already told him, John looked at Neva who returned his gaze with a kind, motherly smile. She said nothing as her dark eyes glistened and John was drawn into them.

As Neva held John's eyes with hers, they sat quietly for a while. John's thoughts raced through eons of time when Neanderthals, *Homo sapiens* and who knows what other hominid species lived on the earth, together. *"Did they have families as we think of them? Did these different species live together in communities to share food and shelter? Did they know, and were they wary of other human species who looked and acted differently? Did they look at a sunset and wonder why they was here and how everything came to be? Did they even think? Did each species believe they were a special part of creation?"* The questions in John's mind were endless and relentless, creating a thirst for John to learn more.

John finally broke the silence and spoke, "This sort of puts a damper on humans being created in God's image doesn't it. What if Neanderthals were the ones made in God's image and we killed them off! Either all this science is wrong or the Bible has a lot of explaining to do!"

"Why John, are you religious?" Neva asked and a warm smile displaying perfect white teeth lit up her face.

"Not really. Neither Lucy nor I have been to church in years. We just got out of the habit and don't really seem to have the time any more. I don't think that religion has a great deal to offer us anyway." John paused, then continued somewhat apologetically, "…but, that doesn't mean I don't believe in God. Even before I talked to you and Sam, I didn't believe that the story of creation in the Bible was literal. The Bible talks about man being made in God's image. It does seem that man is special since we are so far advanced from all other animals. After all, you don't see any apes riding around in little go-carts they've cobbled together out of sticks, vines and rocks like the Flintstones! But seriously, I had no idea that there were several species of humans living on earth at the same time. It seems to me that this raises a lot of questions about man having a premier place in creation and a special relationship with God?"

Neva nodded her head indicating that she understood what John was driving at, then she spoke. "Science can provide a lot factual data about the biological and molecular mechanisms of evolution, when different species lived on earth, for how long, and even what may have caused species to change over time or die out and become extinct."

"But, what science cannot say anything about is whether or not there is a creator. The writings, creeds, myths and superstitions upon which modern religions are based place man in a special place within creation and, as a result, in a special relationship with the creator. In fact, modern religions claim as fact that man has a unique relationship with their God and they teach as well that God takes a special interest in the daily life, health, welfare and afterlife of human beings. Obviously, there is no scientific basis for such beliefs nor are these beliefs testable in technical or scientific terms. They are, quite simply, beliefs, nothing more and nothing less. Nevertheless, these beliefs often stand in conflict with science, creating quite a conundrum that fuels a great deal of controversy between science and religion."

John shook his head in agreement and gave Neva a look that seemed to say *"go on."*

Neva continued, "It is preposterous to believe, as some do, that the earth is only six thousand years old and that religious texts of today provide a literal and accurate description of the process of creation. Biblical texts were never intended to provide a literal accounting of the age of the universe or any scientific explanation about creation. The fact is that the earliest writings upon which modern day religious texts are based were more than likely intended simply to stimulate thought and to provide a foundation for the meditative exploration of the unknown."

"Creation, the universe, is billions of years old and evolution is real. These are facts, and on this there can be no doubt. Nevertheless, if someone believes that certain holy texts contain the absolute and inerrant word of God, it is perhaps easy to understand how they may feel that science is a grand conspiracy that threatens the religious underpinnings upon which their faith is based. Some fundamentalists believe that science is intentionally perpetuating lies and that it is the religious believer's responsibility to protect the sanctity of the holy books and their God by dismissing scientific evidence that is in conflict with those writings and the dogma associated with a particular faith."

"But," John interrupted. "...isn't there, for example, a lot controversy over some science, such as the methods used to date things. Wouldn't some argue that it's not possible to accurately tell if something is six thousand or six million years old?"

"Not really." Neva said. "Methods of dating rocks, fossils, and the age of stars are actually now quite good. Besides, the same dating methods used to determine the age of a fossil that is hundreds of millions of years old has also been used to accurately determine the age of certain biblical artifacts such as the Dead Sea Scrolls, for instance. It has been determined using scientific methods of dating objects that these scrolls were written between 150 BC and 70 AD, dates that religious fundamentalists would absolutely agree upon. Fundamentalists don't seem to have a problem with that. They only have a problem with dates that are more than six thousand years old."

"Religion has been wrong in the past." Neva continued. "In the late 1400's, for example, Christians believed, based on scripture, that the earth was immovable and sat at the center of the universe. Then in the early 1500's in direct conflict with Catholic teaching, Copernicus proposed that the Bible was wrong and that earth revolved around the sun. Copernicus' theory was later confirmed in the early 1600's by Galileo who demonstrated unequivocally that the earth in fact, was not the center of the universe. Both Copernicus and Galileo were persecuted by the Catholic church for their theories and Galileo was tried and convicted of heresy in 1633. Although the Catholic church continued to maintain that the Bible was correct the world moved on and within a few decades, nearly everyone came to know that in truth, the earth orbited around the sun. Nevertheless, it was not until the mid 1800's that Galileo's writings were removed from the Catholic church's list of banned books. Amazingly, it was not until 1992 that Pope John Paul II officially declared that the church was wrong to have persecuted Copernicus and Galileo. More than four hundred years for the church to admit that it, and the biblical texts upon which the church was built, were in error concerning the position of the earth relative to the sun!"

"It may seem that a few thousand years is a long time but remember, religions today are based on teachings and writings that by geological and evolutionary standards are nothing but a moment in time. The New Testament was largely written 50 - 150 years after Christ. The Torah or Old Testament was written over a period of several hundred years roughly 3000 years ago and the Qur'an was written about 1300 years ago. These may seem ancient but several thousand years is literally a drop in the ocean compared to the four billion years that life has existed on earth or the few million years that humans have been evolving along their branch of the tree of life."

"Religious texts were written well before there was any of the scientific understanding that is available today. As a result, there is simply no reason to believe that those texts should provide a technically correct description of the details of creation or the age of the universe. Nor should these holy texts provide an accurate insight into the wonders of evolution that has been taking place over countless eons of time."

"The holy texts upon which modern religions are based were authored by man based on the knowledge and beliefs of the time in which they were written. These writings were perhaps intended solely to provoke contemplative thought for ancient civilizations, providing simple, allegorical stories that addressed a most human question…that is, how did all this come to be?" Neva gestured once again with her hands up as if she were gently, and with great reverence, holding all of creation in the palms of her delicate hands.

"You and I have been talking for some time about the wonderfully elegant and powerful mechanisms that maintain creation and allow life to continue. It is life itself that is a powerful creative force. The fundamental elements of life are precisely tuned to allow creation to continue endlessly. If religion's holy texts are indeed the actual or inspired word of the creator, wouldn't you think that God Almighty would at least have provided in those texts some small hint as to the miracle of DNA? Wouldn't the creator want us to know that within this single molecule lies the roots of a tree capable of endless diversity in form and function, linking all living things past, present and future through a common creative force. All should marvel at the universal simplicity and creative genius of DNA."

Neva stopped and she and John sat for a while in contemplative silence.

CHAPTER 20
Endless Beauty

After some time, John broke the silence. "So, I guess I can understand that life began on earth billions of years ago, probably as a very simple organism. Over hundreds of millions, even billions of years, life evolved into many different types of species…even ultimately giving rise to man. It also makes sense to me that there is little reason why any Bible should describe these processes in any technical or scientific detail since at the time they were written, the authors simply didn't have any knowledge of DNA nor did they understand the mechanism of evolution."

John continued, "I've never thought about evolution much, certainly if I did I didn't think about it in terms of a bush or tree. Quite frankly, I'm stunned to know that most of the species that have ever lived on earth are extinct today. I don't remember much detail from science in school but weren't some species like the dinosaurs supposedly around for an awfully long time, like millions of years? This is much longer than man has been on earth. How could all this just happen? All these different species; where did they come from and where did they all go? It's hard for me to think about how evolution could lead to such change and diversity even over millions and millions of years." John paused and looked toward Neva for answers.

Neva listened quietly as John talked and pondered out loud the many things that were on his mind. When John was done talking, Neva sat for a moment before speaking. "You've a lot of questions John. If you want, I can try to answer some of these but you must remember that much of this information is still emerging. Everyday we learn something new and wonderful about our past and how this marvelous engine of life works. In this sense, just like life itself, knowledge about life continues to evolve."

John was fully and hopelessly captivated by this gentle woman with her piercing dark eyes, quiet voice, and the passion she had for creation. So, not wishing to miss an opportunity to learn more, he asked that she go on.

Neva reminded John that since life first appeared on earth some four billion years ago, it evolved very slowly at first. Indeed, for nearly three billion years life on planet earth existed only as single cells, endless varieties of one cell organisms sloshing around in ageless oceans. In fact for most of its existence, earth has been inhabited primarily by pond scum. Countless single celled species came and went over eons of time, changing very little for hundreds of millions of years. Trillions upon trillions of delicate microfossils can be identified in rock strata dating back three and a half million years. Initially, the lack of these types of structures in the fossil record confused Darwin. His confusion is not surprising since no one at Darwin's time had thought of looking for microscopic fossils. In truth, delicate single cells rarely leave a fossil record. However, once scientists began to look in the right types of rocks using microscopes in the 1940's and 50's, the field of micropaleontology literally exploded as a dizzying array of single cell microfossils were discovered. Cores taken from deep sea sediments provide millions of years of unbroken micofossil evolutionary history detailing changes in single cell organisms which by the trillions upon trillions lived and died, their microscopic carcasses carpeting the ocean floors for countless millennia.

So it was for three billion years. The only inhabitants of this minuscule planet were no more exciting than single cells as earth circled the sun. That sun, but one of one hundred billion stars in one of a hundred billion galaxies peppered throughout a vast and endless universe. Compared to today, oxygen levels in earth's atmosphere three billion years ago remained relatively low and as a result, single celled organisms were the only life the oceans could support. Then suddenly, at least in geological time, about 600 million years ago, the fossil record shows multi-cellular life beginning to appear. These early complex organisms consisted of what we might recognize today as jellyfish, worms and other disgusting creatures that lived during what is now referred to as the pre-Cambrian and early Cambrian periods. For more than 50 million years these soft bodied creatures flourished in the oceans and constituted the only form of complex life on earth. Then, sometime a little more than 500 million years ago organisms began to build shells. Building shells, or mineralization, represents a highly significant advance in development and it occurred at a time when oxygen levels in earth's atmosphere had risen to a level that allowed the process of mineralization to occur. While all life still remained in the sea, the emergence of species with shells signaled that earth's young atmosphere was finally primed to begin to support life on land.

Sometime around 440 million years ago a new branch formed on the tree of life that would ultimately give rise to modern insects and other invertebrates, that is, animals without bones. The timing of this dramatic branching of animal life coincided with a time that the first plant life began to appear on land. The most ancient invertebrate branches include not only the single cell organisms but also multi-cellular organisms that remained tied to the oceans such as sponges, mollusks, clams, snails and the like. These invertebrate species have continued to evolve for more than 400 million years and exist still today in one form or another, bearing a striking resemblance to their ancient descendants. A good example that John was familiar with from his trips to the beach is the horseshoe crab whose ancestral line can be traced through an excellent fossil record all the way back to the Cambrian period over 400 million years ago.

The vast numbers and seemingly limitless variations of invertebrates are astounding to say the least. In fact, 99 percent of all the organisms that inhabit earth today are invertebrates. Even more striking to John was the revelation that if one could somehow weigh the total biomass on earth, that is, the weight of all living matter on the planet including plants and animals, 99 percent of the total weight would be attributable solely to single cell organisms and invertebrates, basically germs and bugs.

There is a story, Neva told John, about a famous British biologist named John Burden Sanderson Haldane (1892 - 1964). During an interview in his elder years, Haldane was asked what he might conclude about the creator based on his, Haldane's vast knowledge of biology. After some quiet reflection, Haldane is said to have answered that the creator must have had an inordinate fondness for beetles. Indeed, beetles represent 40 percent of all the insects on earth and there a more than 350,000 different species of beetles. No one really knows how many different species of living organisms currently exist on earth however, estimates run from 10 - 100 million. This means that beetles alone make up 1 - 3 percent of all living creatures on earth. A creator that made everything must indeed have had a fascination with these critters to make so many of them in so many different varieties.

"Perhaps beetles are God's chosen and they will ultimately inherit the earth." John thought to himself.

"Animals with backbones began branching on the evolutionary tree about 350 million years ago." Neva continued, interrupting John's musings about beetles. "...and, some branches, like sharks, go back even further than that.

The branch that gave rise to the dinosaurs arose some 230 million years ago and yes, in answer to your earlier question John, dinosaurs were around for a very, very long time. However, after ruling the earth for nearly 165 million years, dinosaurs simply vanished leaving only a little bud on their evolutionary stem that ultimately grew into a new branch that gave rise to birds. In fact, scientists now believe that dinosaurs might have had feathers as the process of evolution began to morph these giants into more bird-like creatures."

"But…" John chimed in looking quite perplexed as he interrupted Neva's professorial oratory, "…how did all this take place? What is it about DNA that allows these endless changes in diversity to occur? How can beetles, fish, frogs, dinosaurs, birds, lions, tigers, bears and man all come from the original branch of jellyfish and slime mold? And then, how can an entire species that lived for more than 165 million years simply vanish from the face of the earth?"

"Great questions." Neva smiled. She could see the curious wonder in John's eyes as he tried to absorb all she was revealing to him. She continued.

Neva reminded John of how immense the periods of time were that they were talking about. Referring back to her earlier discussion about Biblical texts being perhaps 2000 to 3000 years old and how humans believe that 3000 years is a terribly long time, Neva asked John to think about 3000 years as a single lifetime. Suppose that a normal human lifespan was 3000 years. How many generations would it take to span a period of 100 million years back to when the dinosaurs roamed the earth Neva asked John. John correctly estimated more than 30,000 generations. So, if a man could live for 3000 years, the dinosaurs lived more than 30,000 human lifetimes ago. Going back even further to a time when only single cell organisms ruled the earth you'd be looking at a million lifetimes ago…and this was if a normal human's lifespan was 3000 years! Taken in this context, 3000 years is not long at all, nor is 30,000 or even 3 million years very long in terms of geological and evolutionary time.

Next, Neva talked to John about evolutionary innovation…that is, how new structures and functions appear. This goes back importantly to the non-gene parts of the DNA; those vast regions of DNA between the genes that were once thought to be junk, but which are now known to contain switches, controls and check points for the actual genes themselves. Simply put, because of multi-functionality, redundancy and modularity of the genetic code, the parts in DNA that determine such things as the size of your brain, where your arms grow on your body, how many legs a spider has, and the number and width of stripes on individual zebras are very, very ancient. Evolution doesn't mean that new structures just appear or disappear, or that new species magically bud off from a main branch. No, evolution works with ancient codes and messages embedded in the DNA. These codes have existed for millennia. Nature tinkers with structure and function over millions of years and countless generations. These changes, perhaps in response to environmental stresses, disease, or a changing food supply get passed on from generation to generation. Those changes that offer some advantage are sustained and get continually modified. Those that provide no, or only a short-lived advantage, fade away and may ultimately lead a species down a path to extinction.

Complimenting genetic tinkering using ancient codes embedded in the DNA is the phenomena of mutation. DNA synthesis, while highly elegant and remarkably precise, is not a perfectly running machine. Occasional mistakes are made when DNA is being copied. While cells have mechanisms for finding and repairing most mistakes, even a nearly perfect quality assurance system will allow errors to slip through from time to time, especially when duplication occurs literally trillions upon trillions of times in a single organism. These errors lead to mutations in the genetic code which occur, for example, when an "A" gets mistakenly substituted by a "G". More often than not this results in no or little change in structure or function due to the built-in redundancy of the genetic code and protein synthetic processes. However, occasionally a single mutation in a critical section of DNA may result in a permanent and significant change in a protein that is encoded by a particular gene. Over generations, this change may become perpetuated and ultimately becomes a new tool for evolution to tinker with. The process is unending.

Neva next posed to John the following question to John. "So John, you have one mother and one father. Your mother had a mother and father, and your father had a mother and father who were your grandparents. That's two parents and four grandparents. Each of your grandparents had a mother and father, so you had eight great grandparents. Each of your great grandparents had a mother and father and each of your sixteen great, great grandparents had a mother and father, etcetera. This is your evolutionary bush. So, if you look back over twenty generations of Mitchells, which probably is four hundred years or so, how many people are you directly related to?"

John could only guess and said, "Hmmmm, quite a lot I suppose!"

Neva sat up proudly smiling and said, "More than one million! If you trace your family tree back only twenty generations, or four hundred years, you are directly related to more than one million different people. If you go back as far as thirty generations, a mere six hundred years, you, John Mitchell have over one billion direct-line relatives! Now, think about a mouse that gives rise to a new generation every seventy days, or a fly that produces a new generation every ten days. Imagine how many relatives a mouse or fly would have over just a few hundred years. The fact is…" Neva concluded, "…that every human alive on the planet earth today is the result of literally billions upon billions of human conceptions. Each time a child is conceived half of its DNA comes from the mother and half from the father. With this process happening over essentially billions of generations, each human being is actually a highly genetically modified organism."

"Wow!" Said John shaking his head and running a hand through his thick hair. "I've never thought of it like that. I guess that means there is a pretty good chance that everyone on earth is related in one way or another to everyone else since it's likely that some of those billions of people crossed paths at one time or another over six hundred years. But, what does that have to do with evolution?"

Neva continued to explain to John that all this mixing of DNA is a critical component of evolution. Over hundreds of millions of years, the mixing of trillions of generations of DNA, coupled with nature's tinkering with structure and function, can result in striking changes in a species, or even cause the branching off of an entirely new species.

It's a humbling thought for any individual to realize that they have literally billions of direct ancestors going back only to the time that Columbus set out from the coast of Spain in search of a westerly route to Asia across the Atlantic. What would anyone suppose are the chances that over the millions of generations since *Homo sapiens* first appeared on earth 200 thousand years ago, that a Caucasian does not have some African American DNA, or an Asian an Indian, a Christian a Jew or, a Muslim a Hindu. Once the reality of DNA and creation are understood it is not a great leap of faith at all to understand that all human beings are indeed, brothers and sisters.

Next, Neva talked to John about the other side of the evolutionary coin, mass extinctions. A mass extinction is when an entire species or collection of species are simply wiped off the face of the planet resulting in the permanent end of an evolutionary branch. Such an event happened with the dinosaurs.

John had heard about theories of a large meteor slamming into earth and wiping out the dinosaurs. It is true that evidence points to a great cataclysmic event on earth around sixty-five million years ago that corresponds roughly with the time of extinction for the dinosaurs. However, there is now considerable debate over whether or not a meteor strike was the reason for the demise of dinosaurs, or whether a meteor just provided the finishing touches on a process that involved disease and extreme climate change that were already leading the dinosaurs down a path to extinction.

To understand a bit of the controversy it is important to know that very clear evidence exists in the geologic record for a meteor strike somewhere in the Gulf of Mexico around sixty-five million years ago. The massive explosion resulting from this collision covered the planet in a layer of dust that is clearly evident in samples of sedimentary rock around the globe. This layer of sedimentary dust is referred to as the impact layer in geological terms. The impact layer can be precisely dated and contains material that originated in the Gulf of Mexico. Scientists know this because the impact layer could only have been deposited on the opposite side of the globe if it was thrown into the atmosphere by a massive explosion and subsequently spread by climatic winds to completely envelop the earth. Indeed, so devastating was this event that it took three million years for the oceans to recover. The controversy, however, revolves around the curious fact that in the layer of dust that marks this event there is not one single dinosaur fossil to be found…not one. At the time of the dinosaurs the land was literally crawling with the beasts as they indeed did rule the earth for more than 160 million years. As some scientists have suggested, if a meteor strike wiped out the dinosaurs in a matter of months following a collision with earth, the ground should have been littered with dinosaur remains piled on top of dinosaur remains. One would think that even under the worst of circumstances some of these dinosaur carcasses should have been preserved as fossils. The lack of a fossil record for such a rapid mass extinction argues for a more prolonged and insidious type of species demise such as might occur through disease or progressive climate change.

"The point…" Neva explained to John, "…is that something happened long ago that changed the earth dramatically and made conditions harsh for a long period of time. Many species died out while others found a way to survive and continued to evolve, coming out of the period as the newly dominant life forms. You see, evolution is perpetual. As long as life can eek out even the barest finger-hold in an environment, life will endure and change. After the dinosaurs, the path was cleared for mammals and birds to begin to evolve and flourish. On the other hand, insects and other invertebrates just seem to keep going and going. Probably due largely to their short generation time invertebrates are able to adapt more rapidly to dramatic ecological changes."

"Also, this was not the only mass extinction event in the history of earth." Neva continued. "The fossil and geological records reveal clear evidence of a total of five major mass extinctions over the four billion years of life's history on earth. And, there will likely be many more to come. We know, based on the past, that the average duration for most mammalian species is several million years. Therefore, there is little reason to expect that man or any other species on earth today will not become extinct after a few million years. The good news is that modern man has only been around for a little more than 100 thousand years so statistically speaking, man has some time to go. That is unless *Homo sapiens* cause their own extinction prematurely."

Neva chuckled and with a comically dismissive wave of her hand, she added, "Regardless of how long it takes man or other species to become extinct, all of this will end about five billion years from now when, at last, the sun will have lived out it's useful life. That's when the sun will explode and give birth to a brand new star and perhaps a new solar system. Then if the conditions are just right, the circle of life just might begin all over again!"

"Oh now there's a cheery thought." John said throwing up both hands in a mockingly helpless gesture.

Then getting serious again, John asked. "But ok, I guess now I can see how evolution can result in amazing diversity over impossibly long periods of time and how species can just disappear due to any of a number of reasons. However, it seems that humans are special since in so many ways we are far advanced compared to other animals. In fact, we're so advanced we could help mother nature along and cause our own extinction! How could we have just developed from random chance and evolutionary tinkering with DNA? Regardless of what you think about God, religion or the Bible, humans are the only species that have the time and the brains to sit around wasting time wondering why we are here and how we came to be. We have big brains and opposable thumbs. We create art, music, literature and great cities. We've even been to the moon. I don't see how evolution could have given us all that while our supposedly closest relatives are still swinging from vines in trees."

Neva thought for a moment before answering. "Remember John, first of all, man did not evolve from monkeys. Rather, all primates share a common ancestor and the primate branch, of which we are a part, budded off from other mammals more than eighty-five million years ago. Hominids or humans separated from our closest primate relative around five million years ago so there have been many, many generations over eons for evolution to get things right…that is, if you wish to assume that modern man is the peak of perfection."

Since monkeys and man share a common ancestor, this explains why humans and monkeys share so much of their DNA. Out of all the species of monkeys around today, humans are most closely related to chimpanzees. There is less than a 3 percent difference between a chimpanzee's DNA and man's. In fact, human DNA compares more closely to the DNA of chimpanzees than the DNA of two different species of frogs compares to each other! Imagine that; according to our DNA man is a closer genetic relative to the chimpanzee than a garden toad is to a bull frog. Even more surprising perhaps, chimp DNA is a closer match to human DNA than chimp DNA is to gorilla DNA. This fact has caused some to wonder if chimpanzees should not be classified in evolutionary terms with humans, and not monkeys. It has taken millions of years and countless generations for the significant changes to take place that make *Homo sapiens* uniquely human, and chimps uniquely chimps.

However, human's link to the past goes even deeper than the DNA that man shares with other creatures. The miracle that is embedded within DNA is common to all species and binds every living creature to one another. There are many examples of identical genes that serve different functions in different species. It is not actually the differences in genes that make humans different from monkeys, mice or birds, but it is the control of genes that differentiates one species from another. There are simply no uniquely "human" genes to be found.

A wonderful example of this which Darwin himself would have found fascinating is a gene called FOXP2. Certainly perhaps one of the most distinguishing characteristics of humans is speech. Man's ability to develop sophisticated languages and communicate with one another is considered one of the pivotal evolutionary advances that has allowed humans to accomplish so much compared to other animal species. In fact, one of the things that may have led to the demise of the Neanderthals was perhaps their inability to communicate effectively and share complex thoughts.

The FOXP2 gene was first discovered in a London family where offspring spanning multiple generations all exhibited a specific inherited speech disorder that prohibited them from speaking intelligibly. Quite simply, family members with the disorder could not put words together to form anything remotely resembling a coherent thought. Neither could affected family members utter a sentence that made any sense what-so-ever. Genetic analysis of the family discovered that affected individuals all had an identical mutated form of a gene in which one single base pair was different from the gene found in family members who were able to speak normally. The gene was named FOXP2 since it was determined to be part of an important group of genes, called FOX, that were known to be critical for normal development.

FOXP2 is a relatively large gene comprised of a nucleic acid sequence nearly 280,000 nucleotides long. And yet, a change in just one of those nucleotides was responsible for the language defect. At first, since FOXP2 seemed to be so important for the development of speech, its discovery was thought to be the genetic holy grail and represented the first uniquely human gene. However, this hypothesis turned out to be wrong and the truth is that FOXP2 is not unique to humans at all. Indeed, FOXP2 can be found in many different species including Neanderthals, monkeys, mice, birds, and even alligators. The gene is highly conserved across all species, meaning that the sequence of nucleic acids in the FOXP2 gene is virtually identical from one species to the next. Amazingly, the FOXP2 gene is critical not only for the development of normal language in humans but, it is responsible as well for the development of song in birds. Birds with a mutated FOXP2 gene cannot attract a mate because they cannot learn how to sing. FOXP2 is also directly involved in the development of ultrasonic vocalizations in mice that are essential for maternal recognition of newborn pups. Mice born with a mutated FOXP2 gene die of starvation since they cannot call out to their mother when they are hungry.

It would appear therefore that FOXP2 is an essential component of normal learning and memory. Indeed, FOXP2 is critical in the formation of sounds in a number of different species whether those sounds are language in humans, songs in birds, or the cries of "mommy" in infant mice. Monkeys, including chimpanzees and gorillas, possess a version of FOXP2 that is very similar to the mutated gene in humans that causes the language defects observed in the London family that led to FOXP2's discovery. For whatever reason, the evolutionary branch of monkeys has maintained the mutated FOXP2 gene whereas on the human evolutionary branch, a change of a single nucleotide in FOXP2 allowed Homo sapiens to develop the ability to communicate through speech.

While humans and monkeys have many of their genes in common, the difference between the species is not because humans have uniquely human genes that are not found in primates. The endless process of evolution over millions upon millions of generations has used ageless control codes embedded in DNA to tinker with genes like FOXP2 and others to alter the micro-anatomy and circuitry of the human brain. Time and experimentation have allowed evolution to modulate the number and types of neuronal connections that ultimately endow man with highly specialized cognitive functions and thought processes. When combined, all these small changes in how and when genes get turned on and turned off during development are the reason why a human embryo becomes a human and a monkey embryo becomes a monkey.

The evolutionary branch that includes modern humans is very bushy. That is to say there has been a great deal of trial and error along the evolutionary path that has resulted in modern humans. In terms of brain size for example, our friend the Neanderthal actually had a much larger brain than modern-day humans. Does this mean Neanderthals were smarter than *Homo sapiens*? Not necessarily. This does mean, however, that brain size alone is not a singularly important determinant of evolutionary success. Many different species of hominid have come and gone over millions of years. Each of these hominid species were distinct from any monkey species alive today, including chimps. Of all the different species of hominid that existed over these millions of years, *Homo sapiens* are only the most recent. Even though humans coexisted on earth for some time with several other hominid species, many species of hominids had existed on earth well before *Homo sapiens* appeared. Many of these hominids species had a far longer tenure on earth than *Homo sapiens* have enjoyed so far. In that regard, *Homo sapiens* are relative newcomers to the planet compared to other hominids that came before.

"On this there is now doubt…" Neva concluded. "…man is but a moment in an ageless and continuous process that has been going on for hundreds of millions of years. The process of evolution is fueled by the beautiful and elegant properties of DNA that enable the reuse and refinement of ancient genetic codes that program biological structure and function. One only needs to look at an early stage human embryo to see that it is virtually indistinguishable from the early stage embryo of a fish, salamander, chicken, rabbit, cow, dog, monkey or nearly any other creature on earth. The fact is that all living creatures, man included, share more in common than they have differences. Man is part of an endless process derived from changes in the genetic code that have occurred over hundreds of millions of years. The process of human evolution is not over. *Homo sapiens* are likely destined to change for millions of years into the future. Evolution does not discriminate or favor one species over another and most importantly, evolution does not stop, ever."

Neva looked directly into the setting sun without turning to John or letting him respond and added with a shake of her head, "I can't understand why some feel it is so important to be…" she used her hands and gestured with quotation marks, "God's chosen". "There is nothing so arrogant than a belief that man is somehow more noble and different from the rest of creation. It is equally incomprehensible to me that despite overwhelming evidence there are those who continue the vacuous notion that creation is only a few thousand years old and that life was created precisely as it exists today. The fact is, all life is sacred and beautiful. Creation is ancient and it is as elegantly complex as it is highly improbable. We have no idea why DNA has been mutating and organisms have been adapting and changing ever since the first cells mysteriously appeared on earth four billion years ago. Life has been evolving from the beginning and it will keep on evolving for billions of years to come, long after man is gone and the religions of today have been long forgotten. No one knows for what purpose different species have come into existence and then disappeared. Assuming, that is, there even is a purpose. The God described in man's holy texts is supposedly infinite and beyond the mind of man. Why then limit such a power to the one defined in primitive writings only a few thousand years old? Why place our deepest beliefs in a personal God revealed through man's feeble and often poorly written musings? Shouldn't the creator of a universe that is an endless number of years old containing countless numbers of galaxies composed of an infinite number of stars spread over impossible distances, stand infinitely beyond our knowing and understanding? What the creator did is right in front of us. All we need to do is look around. When you look and understand just a little of how creation works, one cannot help but to stand in awe."

John and Neva sat at the picnic table in silence enjoying the sunset. Suddenly and without warning the peaceful calm of a quite Virginia evening was shattered when a horn sounded. Both John and Neva to look up simultaneously in the direction of the noise and saw a tow truck pulling up behind Neva's Buick. Fred's Towing Service was emblazoned on the truck's door.

A short stocky man opened the door of the tow truck and jumped out, calling to John and Neva as he did, "Someone call for a tow?"

"God." John said looking at his watch. "We've been talking for more than an hour Neva. Let's see if this guy can help you out."

John and Neva got up from the picnic table and waved to the tow truck driver as they began walking casually toward the parking lot. The sun dipped below the horizon and the sky began to darken as early evening descended like a transparent vale of sheer black fabric.

CHAPTER 21

John and Neva Say Goodbye

"Busy night." Said the short man shaking his head while introducing himself to John and Neva as Milo from Fred's Towing Service. Milo looked to be at least sixty years old, barely five feet tall and nearly as round. A full gray and unruly beard cascaded over a barrel chest and stopped at the top of his substantial belly. Milo looked like a hobbit. His short round stature was in conflict with Milo's high squeaky voice that seemed out of place and would have more suited to the body of taller, scrawny man. John and Neva gave one another a quick smile as they introduced themselves to Milo.

Milo saw immediately the flat tire on the Buick. Taking a look at John and Neva, Milo immediately turned his attention to Neva and observed, "looks like ya got a flat young lady, and a bad one at that. I kin't fix it cuz yer goin' to need a new tire. Thisun' here's shot."

John told Milo that Neva had a spare but it was very low on air.

"Well let's have a look." Milo squeaked. "If all she needs is air I kin fill her up and get ya on yer way fur sure. But ya'll need to get a new tire first thing cuz ya don't want to be drivin' around with no spare."

With that, Milo was in the Buick's trunk in a flash, Milo's feet were nearly off the ground as he reached deep into the trunk for the spare tire making giving him the appearance of a penguin struggling to hoist itself out of the water and onto the ice. Milo wrestled the spare out of the trunk and carried it effortlessly to the truck where he filled the depleted spare with air from the truck's compressor.

"Hope this works." Said Milo. "Sometimes an old tire been low for a long time won't hold air too good. I'd hate for ya to git down the road and go flat right away." He added.

"That wouldn't be good." John remarked looking at Neva and shaking his head. "It might be better just to get Neva to a place where she can get a new tire."

Milo commented that the spare seemed to be holding air ok but he thought he heard a small hiss coming from somewhere around the rim and he agreed that it might not be worth the chance. Scratching his beard for a moment, his eyes suddenly lit up. "There's a National Tire and Battery shop not far off the next exit on the way back to Fred's. I could drop ya there sure'nuf. I think NTB's open til nine or even later."

"Ain't goin' to be much more expensive for me to tow ya a few miles than to change the tire ma'am?" Milo said with a shrug of his shoulders looking at Neva, his expression asking her what she'd like to do.

Neva turned to John, then to Milo and she said that she'd feel much better if she just got a new tire but she wondered what she would do if the NTB wasn't open.

"OK, le'me check with dispatch and see if they know how late NTB is open. If it is, I kin get ya rigged up and we kin git goin' in a jiffy." Milo said with a nod of his head indicating that he agreed this was the best thing to do. Milo called Fred's dispatch to tell them what he was doing and they confirmed that the NTB store was in fact open until ten o'clock that evening. They should have plenty of time since it was now barely 7:30.

"Ya kin ride with me." Milo said as he busied himself putting the spare back into Neva's trunk, dragging chains off his truck and getting things ready to tow Neva to the NTB. "We'll git ya fixed up and on yer way in no time." he added.

John and Neva backed away in the direction of John's car so as to not be in Milo's way. As they watched Milo work John looked at Neva and said with a hint of skepticism and concern in his voice, "Perhaps you should ride with me and I'll follow the tow to the NTB just to make sure that everything is all right."

"Oh don't worry about me John." Neva said with a smile. "You've been more than kind already and I don't want to delay you any longer. Really, I'll be fine. I'm used to being on my own."

Neva lowered her voice and leaned in closer to John and whispered, "Milo seems like a fine little elf. I'm sure he'll take good care of me."

John shook his head, "Lucy would be upset with me if she knew I left you on your own."

Neva smiled and patted John softly on the arm. "Now John, you have been quite gallant this evening helping me out and I will be perfectly fine. You need to get on the road as you have a long drive still ahead of you. I can't tell you how grateful I am and much I've appreciate your staying with me until the tow truck got here. You were great company and I enjoyed our talking with one another."

Just then, Milo squeaked, "All sit! Ya ready to git on the road young lady?"

Before John could protest further Neva had turned toward him and placed her hands gently on each of John's shoulders. Looking up Neva peered deep into John's eyes. Her dark eyes sparkled as her gaze seemed to bore deep into John's mind.

"Thanks again John, it was a pleasure meeting you." Neva said. Then quite unexpectedly Neva leaned toward John while stretching up onto her toes and gave him a soft, warm kiss on his cheek.

"But…" John began, and before he could continue Neva pulled away and turned toward the tow truck. As she walked away Neva gave John a final parting glance over her shoulder and said with a smile, "Take care John. Drive carefully and again, thanks. Have a wonderful life."

John watched in silence as Milo opened the passenger door for Neva and quickly cleared off an arm full of rags, papers and a couple well-used coffee mugs, shoving them into the middle of the wide bench seat of the tow truck. Neva turned and smiled once again at John as she climbed into the truck. Milo closed the door and with a nod gave John a friendly wave as he walked to the driver's side of the truck and climbed in.

As John stood by his car and the tow truck pulled away, Milo and Neva both gave him a final parting wave.

"I hope that guy is ok." John thought to himself.

John stood with his hand on the door handle of his BMW just staring at the picnic table where he and Neva been sitting. *"She was really a nice lady. And smart too."* John thought smiling while continuing to gaze at the now empty table in the rest area. Then shaking his head he said out loud, "Boy, what an interesting stop this has been."

John began to replay in his mind bits and pieces of the fascinating conversations he had had with Sam and Neva. Light years, empty holes in space, parallel universes, unfathomable periods of time, Big Bangs, DNA, evolution, extinct species… It was then that John cocked his head, screwed up his eyebrows and said out loud although there was no one to hear, "Imagine that, most everything that has ever lived on earth is now extinct. Wonder what will happen to us in ten million years?"

John shook his head and listened, though he knew not for what. No one answered. Minutes passed. Finally, John shrugged his shoulders, took a deep breath, opened his car and got in. He started the engine and backed out of his parking space. Just as he did, John noticed in his peripheral vision two cars followed by a semi lumbering up the ramp into the rest area. The semi turned left into the section designated for trucks and the two cars veered right toward the cars only area where John was. As they parked, three more cars were also coming off the highway into the rest area.

"*Hmmmm.*" John thought to himself. "*All this time and not a soul here. Now the place is jumping. Go figure.*"

John accelerated quickly down the ramp from rest area and onto the highway entry lane. Checking his mirrors for oncoming traffic John merged into the right lane and smoothly jumped to seventy miles per hour, once again heading south on I-95 toward Richmond. In just a few minutes, John was at the next exit. As he passed the exit ramp he saw the tow truck with Neva's car behind it. Milo was at the end of the ramp and beginning to turn left onto the road that passed over the interstate. Even though it was getting dark John was sure he could see Milo and Neva chatting in the front seat of the truck, both smiling. John beeped his horn as he zoomed under the overpass and the tow truck lumbered, seemingly in slow motion, onto the overpass directly above.

CHAPTER 22
Traveling at the Speed of Light

The bright light that had enveloped him from all directions began to fade. He felt no pain. In fact, he felt nothing at all and seemed to be weightless. In a surreal dream state he was vaguely aware of things around him, but nothing had any real form or substance. He could hear sounds, a deep rumble, not unlike the roar you can hear when you put your fingers in your ears and listen to the engine of life throb deep within your skull.

The ambulance arrived about ten minutes after the state trooper had gotten to the accident scene. Before the ambulance appeared there had been nothing anyone could do for the injured driver other than to keep him warm with a blanket. A woman in her mid thirties knelt beside the driver and held his hand while her other hand rested gently on his shoulder, patting gently him from time to time. She spoke softly telling him that help was on the way.

"Just hold on, everything will be alright." She kept repeating.

There was a good deal of blood on the injured driver's head and face from a deep cut on his scalp upon which the trooper had taped a large packet of gauze from the emergency kit the trooper carried in his cruiser. The bleeding from the gash in the injured man's head had mostly stopped, but it was clear that the man had other, unseen injuries. His face was badly bruised and covered with numerous small cuts, presumably from flying glass. His abdomen was becoming severely swollen and his breathing labored and shallow.

The EMT's were quick to respond as soon as the ambulance arrived. They asked how the injured man had gotten out of the car. Earl explained what he and the other men had done to get the driver out of the car.

"We tried to be as gentle as we could, but we thought getting him out fast was more important." Earl said. "Do you think he'll make it?" Earl asked in a concerned voice.

"We'll do everything we can." Said one of the EMT's who was checking the injured driver's pupillary response and airway while the other EMT took stock of the mans arms, chest, abdomen and legs. "You did the right thing to get him out of that car, buddy. It would have been much worse for him to hang upside down in there until we got here or even worse to be in there if the car caught fire. Nice work." Said the EMT, his attention never leaving the injured driver. Then turning ever so briefly toward Earl he added, "You're a hero buddy, thanks."

The EMT's worked quickly to stabilize the injured driver's head and neck then carefully transferred him to the solid transport board. After attaching an oxygen mask, the EMT spoke slowly and distinctly to the injured driver.

"We are going to help you and get you to a hospital. Can you hear me?"

The injured man moaned but otherwise provided no meaningful response.

Without any prompting, the trooper instinctively moved curious people out of the way as the EMT's placed the injured man onto a stretcher and prepared to move him into the back of the ambulance. Fredericksburg had the nearest level one trauma center and was less than twenty minutes north of the accident site assuming the ambulance could get onto the northbound part of I-95 quickly. Fortunately, about a half mile north of the start of the construction zone there was an emergency vehicle crossover they could access.

By now traffic was in gridlock for many miles north of the accident on southbound I-95. Northbound was not much better as traffic slowed and northbound drivers curiously craned their necks to see what was going on.

Once the patient was loaded into the back of the ambulance the trooper and ambulance turned around and headed northbound on the narrow berm until they reached the emergency crossover. The trooper jumped out his cruiser and began directing cars out of the way and off to the side of the road so that the ambulance could get across traffic and onto I-95 north. In just a few minutes the ambulance was roaring north, its siren blaring. The trooper drove back to the accident site to wait for the tow truck which he had learned was on its way and should arrive soon. It would be some time before the southbound lanes of I-95 would be cleared and normal traffic could resume.

The injured driver had the sensation that he was traveling but with no apparent destination or purposeful direction. At the same time, while he had no idea where he was going or how fast he was traveling, his mind screamed that he had to get there fast, wherever "there" was. He began to feel anxious and there seemed to be a great weight crushing his chest making it difficult for him to breath. It was the same feeling someone might get when they are late for an an important appointment and every minute's delay brings increased anxiety. As if in a dream, he was trying to run from something, or perhaps toward something. His arms and legs were made of lead. Every step required tremendous effort. It took an eternity to for him to travel just a few inches. He was going nowhere. Frustration grew. He opened his mouth to scream but could make no sound, his face frozen in a silent scream that no one heard.

CHAPTER 23
Food for Thought

John had driven for about an hour since leaving the rest stop. His progress was becoming slowed by increasingly significant traffic headed south out of Richmond and he began to feel hungry. With more than two hours to go before he reached home, John knew the girls would be in bed by the time he arrived. While he had hoped to surprise his family, Lucy was not expecting him until very late that night so he would be on his own for something to eat. Although the time John had spent with Sam, Neva and the sunset resulted in some delay, time had passed surprisingly quickly. Not anxious at all about the lost time, John was just hungry. A look at the BMW's gas gauge also told him that it would be a good idea to fill up his tank as well. A little food, some gas, a quick stop at a rest room and John would not have to stop again until he was home. Decision made.

John began to look for an exit. John knew, as does anyone who has driven the nation's highways, that there are basically three types of exits when you are looking for gas, something to eat or need to find a restroom. Some exits offer a cornucopia of gas, food and motel options, all within easy off and easy on access to the freeway. These tend to be located near larger metropolitan areas or near the intersection of major highways. The second type of exit for the hungry traveler offers adequate, but fewer options. These tend to be near to the freeway as well, but often can look somewhat desolate and uninviting. Finally, there are the trick exits. These entice you off the highway with the promise of one or more fast food restaurants of your choice, however, you don't realize until you've pulled off the highway and onto the exit ramp that the McDonald's you were looking for is a mile or more from the highway. You know this by the little blue sign on the exit ramp with the McDonald's sign poised over an arrow pointing to the right that says "1.1 mi". Often, this third offering takes you into or through a small town and one of the borough's primary means of extra revenue generation, a speed trap. Local police lie in wait ready to nail the traveler who absolutely had to get off the highway to find a bathroom and decided that it was better to speed to the McDonald's only 1.1 mile from the freeway than to travel 10 miles or more to the next exit on the interstate.

This latter of exit was the type on which John found himself. Three fast food restaurants and two gas stations were promised on the signage leading up to the exit ramp. However, once on the ramp John grunted in disgust to see that Wendy's, the closest eatery, was 0.7 miles east and a BP station was 0.4 miles in the same direction. Hardly convenient. John thought for a moment about getting back on the highway and trying the next exit. Then again, a Frosty at Wendy's sounded pretty good and he really did not feel rushed to get home. Now that he was past Richmond, even under the worst of circumstances he would still arrive home hours earlier than he'd anticipated when the day had begun at 6:00 AM that morning in Washington DC. He would have a bite to eat, call Lucy, then be on his way for the remaining two hour journey home.

Turning off the exit and heading east, John soon passed the BP station on his left. *"Good."* He thought, *"I can eat and then get gas on my way back to the interstate. Shouldn't take long at all."*

After only a few moments he could see the familiar Wendy's sign ahead on his right. As he approached he saw that there were only dim lights in the building and no other cars in the parking lot. He slowed and pulled into the parking lot where he could see more clearly that the indoor lighting was minimal and the Wendy's appeared to be closed. He drove slowly past the entrance door on his left and saw a hand-written sign.

"CLOSED UNTIL NOON TOMORROW. PLUMBING PROBLEMS. SORRY." The sign read.

"Shit." John spit under his breath. "How far were those other fast food joints, a McDonald's or Arby's? Were they even this direction?" John said to no one in particular. Heaving a sigh and shaking his head John came to the quick realization that he'd made a bad decision on the choice of exit.

Gritting his teeth, his stomach now grumbling, John drove around the back of the Wendy's following the exit sign and arrows leading back toward the main road. Hesitating just a bit because he was not quite sure which way to turn, John looked up and noticed for the first time the small diner across the road almost directly opposite from Wendy's.

"Funny..." John thought to himself a bit puzzled, *"...I didn't see that when I pulled into Wendy's. Must be some kind of parallel universe on that side of the road."* John thought, smiling to himself.

John sat at the Wendy's exit for a moment debating his options. He really had no recall as to whether the other fast food joints were east or west of the interstate. If he guessed wrong he would waste more time just driving around. On the other hand, getting something to eat at the diner could take longer than a fast food restaurant. *"Maybe a cup of coffee would be better than a Frosty anyway. Its been a long day."* John concluded to himself.

Just across the road from where John sat in his car, SAM'S DINER was a small, well lit establishment with several vehicles in the parking lot. The diner's shiny chrome exterior gleamed, reflecting light from a roadside lamppost. Three horizontal red stripes adorned the diner's chrome exterior. Two stripes ran the length of the entire structure just below and another above the windows and a third red stripe circled the diner just below the curved roof line. The diner reminded John of a giant Airstream trailer on blocks. A small portico jutted out from the center of the building and served as the entrance. Exterior and interior doors of the portico provided an airlock to the diner itself. Stereotypical for many diners, a neon sign on the roof read "Great Home Cooked Food", written in cursive just below the diner's name which, stood out prominently in large capital block letters. A bright starburst in the upper left corner of the neon sign completed the dazzling light display that would have not been out of place on the Vegas Strip. Inside the brightly lit diner John could see through the windows what appeared to be two couples at booths and a few people who seemed to be sitting at the counter.

"Sam's diner." John thought looking at the sign in near disbelief. *"Now, there is a coincidence. I'll bet they have an cosmic burger on the menu."*

John decided he had time, and besides, he knew from experience that the food at most roadside diners was generally far superior to that from any fast food restaurant and typically not much more expensive. Sam's was obviously not terribly busy and John decided it wouldn't take that long to have a cheese hamburger. John drove across the highway and pulled into Sam's parking lot beside a red Ford pickup truck with a large metal tool box fitted across the truck's bed just behind the cab under the rear window. John's was the only BMW in the parking lot.

As John got out of his car and was entered the diner, a large man who had been sitting on a stool at the counter rose, left a five dollar bill on the counter and was on his way out. The man smiled and nodded politely to John while holding the door so John could enter. John thanked the man and watched as he headed toward the red pickup. Another man was now paying his bill as well, obviously about to leave. The diner's counter was empty. An obese couple was sitting in a booth at the far end near a back window and a young family with a small child in a booster seat occupied a booth nearest to the diner's entrance.

Sam's was a prototypical home cookin' diner with metallic-flecked red vinyl padded booths. Along the counter and bolted to the floor were round stools with no backs covered in the same glistening metallic vinyl upholstery. White formica counter tops speckled with flecks of silver surfaced the ample counter as well as every table in the establishment. The counter and table tops were adorned with glistening chrome ribbed side molding and looked as if they had just recently been waxed and buffed. Every other booth bench seat had a chrome pole attached with three hooks protruding from the top for hanging coats and jackets. Behind the counter next to the pass-through window that obviously led to the kitchen, John could see a variety of what had to be home made pies temptingly on display in a chrome cabinet protected by sliding glass doors. On the counter were two round chrome platters rising 6 inches above the countertop on conical chrome pedestals, each platter covered by a high round plastic top. A chrome ball in the center of each plastic top served as a handle. Each platter was lined with a white paper doily. In one platter, a chocolate layer cake with several slices missing looked moist and delicious. In the other, an assortment of pastries and donuts beckoned, juicy cherries, lemon cream filling and glistening white icing reflected the diner lights and provided an almost irresistible aura of sweet delights. Other than the counter where the two men had just finished their meal and their dirty plates had yet to be bussed, the establishment was immaculate.

A small wiry woman wearing a white knee-length dress and red apron was behind the counter at a cash register. Reading glasses hung from a beaded chain around her neck as the woman gave the man at the counter his change. John immediately caught the woman's eye as he entered the diner with an expectant look and the non-verbal question of "where should I sit?" The woman smiled at John and in a voice husky and sultry from years of smoking said with a friendly smile, "Take a seat wherever you'd like hon. I'll be with you in a just a minute."

Returning her attention again to the man who'd just paid his bill the woman in the red apron spoke in a familiar fashion. "See ya next time Frank. Give my best to Darlene and the kids."

Frank, obviously a regular customer waved while looking distractedly back over his shoulder as he headed for the door, fumbling to get his paper change back into a money clip. "See ya Sam." he said, teeth clenched tightly onto the toothpick he held between his teeth. The woman immediately bussed the dirty dishes from the counter.

"Must be Samantha's diner." John thought as he sat heavily onto one of the stools at the counter. John spun around on the stool and planted both elbows firmly on the gleaming white Formica counter top. John placed his feet comfortably on the black tile step that ran the entire counter length about twelve inches from the floor. *"What are the chances I'd meet a Sam and a Samantha in one day with two stops?"* John pondered the odds in his mind while he extracted a copy of the laminated menu from one of the chrome condiment corrals placed at regular intervals along the back edge of the counter. In addition to several copies of the menu, each chrome corral contained a set of glass salt and pepper shakers (chrome tops of course), real glass bottles of ketchup and mustard, packets of sugar and artificial sweetener, a bottle of tabasco and a bottle of Frank's hot sauce.

The menu was fairly extensive for such a small place. John didn't actually need the menu and was really just looking to have something to do while he waited for Samantha to take his order. He knew before he walked in that he wanted a cheeseburger, fries and a soft drink. In hardly a minute from the time he'd sat down, Samantha appeared in front of John. She was at least seventy and perhaps 5'5" or 5'6" and thin as rail. Her deeply tanned face bore fine wrinkles from years in the sun and was neatly framed by an impressive mane of beautiful dark red hair, obviously dyed since the color of Sam's hair did not exist in nature. Samantha wore no makeup other than bright red lipstick which matched the color of her apron. Except for a few smears and flecks of lipstick on a couple of her front teeth, her smile and perfect teeth were as white as the counter tops. Samantha's eyes were bright blue and sparkled as they reflected light and the shiney chrome trim that adorned every surface in the dinner that was not covered with linoleum, formica or vinyl.

"So sweetie, what can I get ya?" Samantha asked with a pleasant smile and a wink. Her sultry voice suggested perhaps she would be amenable to some innocent flirting if John were interested.

"Cheeseburger, fries and a coke." Answered John very business like, then added. "I heard that fellow call you Sam as he left. Short for Samantha I suppose? Are you the owner?"

"Yes indeed I am." Sam replied proudly. Gesturing over her shoulder toward the back of the diner with her thumb and another wink, Sam continued, "My boyfriend Evan, he's the cook. He and I have been here forever. I'm Samantha but all my friends call me Sam." She said, patting John on the hand, winking again and smiling to let John know he too was welcome to call her Sam.

"I'll put this right in and get your coke. Like some water as well?" Sam said.

John nodded in the affirmative, "Yes, thanks."

Sam turned to the kitchen window, stuck the green and white order slip in one of the clips on the chrome wheel suspended from the center of the pass through window, spun the wheel toward the kitchen and said, "Evan, cheeseburger and fries."

Sam returned in a minute with John's coke and a glass of water. "Food should be up in a jiffy hon. Let me know if you need anything else in the meantime." Sam smiled and winked again at John.

"Actually a very attractive woman for her age." John thought to himself.

The large couple in the corner was getting up to leave and they met Sam at the counter by the cash register where she totaled their bill. John watched the couple pay and leave the diner while he enjoyed his coke and waited for his food to arrive. As he sipped, John couldn't help but think about the day's events and the conversations he'd had earlier with the young Sam and then Neva. He'd learned so many interesting things about the universe, its origin, biology and evolution. His head was swimming with thoughts and many, many questions. *"Sam and Neva…what were the chances?"* John thought to himself, amused by the improbability of his day since pulling into the oddly deserted rest stop on I-95 hours earlier.

CHAPTER 24

Sam, Evan and Cherry Pie

Elbows on the counter and his chin resting on folded hands, John drifted deeper in thought. *"The universe began as pure energy. Billions and billions of stars. Distances that no one could travel in a billion life times. How did life begin? No way that molecules could just randomly bump into one another over time to form even a short piece of DNA. Even if they did, how would the DNA duplicate itself without enzymes and a source of energy and nucleic acids? Four billion years for life to evolve into an infinite number of forms and most of these were dead ends. Several different human species inhabiting earth together for ten's of thousands of years. What will man be like a million years from now? Will we even be around? Did Neva and Sam say something about an infinite number of parallel universes with new ones being created instantly all the time? Is God responsible for all this? Why does anything exist at all?"*

"Offering a word of prayer before your food gets here son?"

John was startled from his thoughts as a white plate with a green boarder slid between his elbows. Instantly he could smell the cheeseburger and fries and John could literally feel their warmth on his face and he snapped back to reality. Standing directly in front of him was an elderly man who appeared to be about Sam's age or perhaps a few years older. He was wearing white pants, a white tee-shirt with its sleeves rolled up exposing tanned wiry arms. A clean white apron was tied tightly around his waste and a white dishtowel with a red stripe was draped over one slender shoulder. The man had not shaved in several days and a spotty, mostly grey stubble framed a pleasant smile with absolutely perfect white teeth.

His bright blue eye's twinkling, the man said, "Sorry son, didn't mean to interrupt your meditation."

"Oh no problem. I was just thinking about…" John paused and thought he'd not bother the man with what he'd been thinking about, and instead changed course abruptly saying, "You must be Evan the cook. This burger looks and smells fantastic. Thanks."

John gazed around quickly and Sam was nowhere to be seen. Also, the young couple and child that had been at the booth near the door were gone as well. John figured he really must have been deep in thought to have missed their exit. The diner was now entirely empty except for John and the old man.

"Where's Sam?" John added with a puzzled look.

"Oh she's back in the galley cleaning up some dishes. And yep, I'm Evan, the cook." The man said straightening to his full height of perhaps five feet seven inches while puffing out his scrawny chest in an obvious display of great pride in his profession.

Evan was roughly the same height as Samantha and just as thin. He sported a full head of unruly grey hair, cut fairly short but not short enough to prevent shocks from sticking out in random directions all over the top of his head. John was immediately struck with the thought that Samantha looked like Evan in drag.

John must have been unconsciously staring at Evan with a puzzled look causing Evan to comment, "I know I'm not as good lookin' as Sam, but I do make quite a fine cheeseburger. You better eat before it gets cold. Sam will be back in just a minute. In the meantime, just give a yell if you need anything." In a flash, Evan turned and vanished into the kitchen.

John shook his head in disbelief. "*What a day.*" He thought. "*I need to eat and get on my way back to Lucy and the real world. This is just weird. Lucy won't believe anything I tell her about this.*"

John shook a healthy pool of ketchup onto his plate for the fries and put some mustard on the burger. He added the fresh lettuce, onion, tomato and pickle slices that were laying on the naked top half of a toasted bun and took a bite. Best cheeseburger he had ever tasted, bar none. He wolfed down the burger as if he'd not eaten in days.

As he was scraping the remaining molecules of ketchup from his plate with the last of the hot, crispy french fries, Sam appeared from the kitchen. "Hated it didn't you!" Sam said with a wink and little chuckle as she looked from John's plate into his eyes, smiling knowingly.

"That was really, really good." John said with a satisfied sigh, stretching back on the counter stool and giving his stomach a little pat.

"How about a warm piece of pie and a cup of coffee to top it off before you go? I can tell you for sure that these pies are the best you've ever had." Sam said, indicating with a turn of her head toward the pie display behind and to her right. "Evan makes them fresh every day from scratch."

"He definitely makes the best cheeseburger I've ever had."

"Oh, he can make anything." Sam said. "And, its all good."

John couldn't help himself. Once again a door was flung open revealing yet another path on this most unusual of days. He was drawn like a moth to a flame. "Sounds almost biblical. Which, is not surprising given the day I've been having." John said shaking his head.

Sam's eyes squinted and her forehead wrinkled. A look of concern drifted over her face as she looked at John. "I'm sorry. Didn't mean to pry. Have you had a tough day hon?" She said, her sultry voice shifting seamlessly from flirtatious to legitimate concern.

"Oh no, nothing bad, nothing bad at all." Said John. "In fact, it's been a wonderful day. Curious in many ways, but wonderful. Earlier today I met two of the most interesting and unusual people. One was named Sam by the way…a boy, just a kid really, and the other person was a wonderful woman named Neva. We talked for hours about all kinds of things. The universe, creation, the Bible, God….well, we never really got to talking about God much but he always seemed to come up in the end. It has got me thinking that's all. Raised a lot of questions in my mind if you know what I mean. And then, when you said that 'Evan can make anything and it was good', it just sort of reminded me of the beginning of the Bible. It's nothing, just commenting." John looked at Sam and gave her his most reassuring smile.

"Oh honey…" Sam's demeanor brightened, no longer concerned. "…if you want to talk about God, you should talk to Evan. I think he's read every good book there is. At one time or another, he's practiced nearly every religion you can think of too. It's kind of his hobby ya know. He'll talk your leg off if you get him going."

John began to shake his head and was about to tell Sam that it was not necessary to bother Evan but before John could get a word out, Sam turned her head over her shoulder and shouted toward the kitchen, "Evan, this young fellow wants to talk to you. Get out here."

"Oh really, that's not necessary at all." John protested with a chuckle holding up his hands and shaking his head even though the cat was already out of the bag so to speak. Quite frankly, John didn't want to get in to another conversation the likes of which he had had with Sam and Neva earlier. He was already mentally exhausted from his day and was now eager to get home. Truly John didn't think his mind could absorb any more information today.

Ignoring his protest completely and before John could say another word, Sam turned to John and said, "You are definitely going to need a piece of pie for this sweetie. What kind would you like…it's on the house? A scoop of ice cream and some coffee to go with it?"

"Well cherry is my favorite, but really, I don't…" John tried to continue to protest.

"Oh don't be silly, it's no trouble at all. Cherry it is." In a flash John was looking down at a beautiful piece of fresh warm cherry pie topped with vanilla ice cream. Tiny streams of ice cream were already beginning to ooze like molten lava over the edges of the pie, dripping onto the plate in lazy ice cream falls from the light flakey pie crust.

Sam set a cup in front of John and poured him a steaming cup of coffee. "Let me know if you need more coffee hon." Then, shaking her head and turning on her heal, Sam headed for the kitchen and disappeared. John heard Sam say, "Evan, for cryin' out loud, this young man out here needs to talk to you. Didn't you hear me the first time? Are you deaf old man?"

Had he not already had two odd experiences today John probably would have left a twenty dollar bill on the counter and ran for his car the moment Sam was out of sight. However, the cherry pie looked fantastic and quite irresistible. Besides, curiosity now had the best of him. John mused to himself as he looked longingly at the pie that, worst case scenario, he would just eat the pie while Evan rambled on then John could excuse himself as needing to get on the road since it had been a long day. Ten minutes tops to finish the pie, drink the coffee and make his apologies. Picking up his fork, John smiled as he cut off a healthy piece of the warm juicy cherry pie from the point of the pie wedge. Scooping up a bit of ice cream along with the pie he took a bite. It was a slice of heaven.

CHAPTER 25
Where am I?

The injured man saw bright lights everywhere which was odd since he was not really sure if his eyes were open or closed. It didn't matter. Was he seeing with his eyes, or was the light merely in his mind? He had no way of knowing since at the moment, his eyes and his mind's eye were one in the same.

Tugging. Struggling. Warm hands touched him with great care and tenderness. He was frightened, then calm. Cold, then warm. Anxious, then at peace. A voice.

"Sir, Sir. Can you hear me? You are in a hospital in Virginia. You've been in an accident. We are going to take care of you but I need you to know if you can hear me. You don't have to try to speak, just blink your eyes if you hear me." The voice commanded.

He could hear and understood, but didn't know if he was blinking his eyes or not. *"In a hospital? Accident? Where?"* It was maddening.

Sounds were all around him. Frantic activity as people talked to each other, to no one. He heard "on three" and he seemed for a moment to float weightlessly through the air from one place to another. He landed with a palpable jolt. Unable to turn his head he tried to move his eyes from side to side, up and down. Lights flickered. He felt his clothes being tugged or torn. He couldn't tell. A calm voice seemed to be issuing orders and other voices responded, all voices speaking at once. The injured man could not understand what was being said. Lights flashed in his eyes and he tried to squint, or was he blinking as instructed. He had no idea. Lights grew brighter and brighter. Then suddenly, it was dark again. Silent and cold. He was alone.

CHAPTER 26

What's on Your Mind Son?

"Sooooo, Sam says you want to talk about God." Evan smiled as he floated from the kitchen to the counter where John was sitting. John looked up with his mouth full of cherry pie, a drop of melted ice cream clinging to his upper lip from his most recent delicious bite. "We'll you've come to the right person." Evan continued.

John signaled that his mouth was full and he brought a napkin to his lips. Not noticing earlier, John was struck by Evan's rich baritone voice. Its melodic quality and deep resonance made Evan sound as if he was chanting inside a great cathedral, enunciating clearly each syllable of a medieval canticle that echoed softly off marble floors, walls, apses and pillars.

Wiping leathery hands on the towel over his shoulder, Evan leaned in close and placed an elbow on the counter opposite from John. He looked directly at John with a wry smile. Before John could swallow his most recent bite and offer a response, Evan continued, "I've been a Muslim, a Jew, a Catholic, a Hindu, a Baptist, Presbyterian, atheist, jainist, a spiritualist, a Baha'i, Shinto, humanist, cultist, a Buddhist although, that's not really a religion you know since they don't believe in God but they are not atheists either. I've been a scientologist, a mormon, a zoroastrian, even a free mason....you name it, I've practiced it. Read most of their propaganda too. I've knelt and prayed to the Lord God Almighty, Yahweh, Allah, Jesus Christ, Shangdi, the Holy Spirit, Hari, Ishvara, Zeus, our Father, he who shall not be named, gnowee, Akua, and my lucky stars. So what do you want to know? Go ahead and ask me anything. I've got an opinion on nearly everything under the sun and then some. Sam will tell ya that." Evan smiled, winked and became silent, his piercing crystal blue eyes holding John's gaze in a clear indication that now was an appropriate time for John to speak.

John thought for a moment. There was a great deal on his mind indeed. He remembered leaving Washington after his meetings had ended earlier in the day. He was excited to be headed home much earlier than he had hoped when his day first began. It had been a wonderful day for a drive. John thought back to an NPR report about a giant hole in the universe. That sunset had been so beautiful. The deserted rest stop. Talking for what seemed like hours with a curious boy who seemed to materialize from nowhere. Another strange chance encounter with Neva. John's mind was filled with wonder about what he'd learned over the past hours about creation, life, the universe. But at the same time, John had developed a deep sense of emptiness, his mind was cluttered, confused and held some disturbing questions for which he had no answers.

John had no idea why he'd run into Sam or Neva, nor did it make any sense that he'd get into the types of discussions with them that he had. Most of all, it was unsettling for him by the way every one of his conversations had eventually turned time and again to God. This was not like him. John just didn't spend time thinking about much of anything except work and his family. Certainly, neither God nor religion ever entered into his daily thoughts. He simply had no time for such things and, truth be told, religion held very little interest. God did not seem relevant in his life. With all the pain and suffering in the world it was impossible for John to devote thoughts to a God who, though supposedly good and caring, seemed to just let things run amok. And yet everything he'd learned from Sam and Neva had brought him face to face with a creator, a power, which from the perspective of creation was somehow different from the God he'd remembered learning about in Sunday school so many years ago. The creative power that was in some way responsible for a massive hole in the universe was not the creator that was described in the Bible. This life force that John had learned so much about over the past hours that powered evolution and gave birth to galaxies was decidedly different from the God that stood as the central figure of the protestant religion John remembered from his childhood. Admittedly his knowledge of religions other than Christianity was practically non-existent and even what he knew about Christianity was meager to say the least. However, in John's mind, all religions seemingly worshiped a God that was clearly different from the one that Sam and Neva had alluded to in response to some of John's questions. In fact, John wasn't even sure Sam or Neva had said anything about God in the way John thought about God. They just talked about creation, or was it the creator? John couldn't remember. In essence, John's mind, his inner voice, felt suddenly and hopelessly lost, frighteningly alone, and thoroughly confused.

Evan's rich baritone voice broke into John's thoughts. "Heat rises but its cold on the mountain, eh?"

John looked up at Evan in surprise. "Excuse me, what did you just say?"

"Heat rises, but its cold on the mountain." Evan repeated, then added, "It's just an expression of mine that can mean a lot of different things like, 'cat got your tongue'. You know, sometimes you have so much running around your mind, things welling up inside, you that you think your head might explode. But at the same time you don't really know what to say, what to ask, or even where to begin. So much to ask and not a word to say. Makes no sense does it? Sort of like the fact that while heat rises, how can it be cold on a mountain top? Makes no sense at all does it? I don't know, I guess I just like the expression."

"It's been a completely strange day for sure." John said shaking his head. "You know you're right. I've a lot of things running around in my mind and I hardly know what to say or even whether I want to say anything." Pausing for a moment, John offered a comment about Evan without even thinking. "You seem quite the expert about God?"

"Not that there is such a thing…an expert on God I mean." Evan began, "But, I've kind of made it my life's hobby you know what I'm saying? I'm interested in what other people believe and why. Did you know there's more than four thousand different religions practiced around the world. Can't all be right I suspect, that is, assuming any of them are. It's hard for people to realize that whatever they believe, most of the rest of the world believes something completely different, and usually just as strongly. Still, people have been going to war over what they believe since the dawn of man. They persecute, judge and kill one other, all in the name of God. Governments and laws are often based around religious beliefs. Some nations and its citizens even think they are favored over other nations in the eyes of God. Earth's been around for billions of years and will be long after we've gone extinct. Seems to me if the creator does prefer one country over another, he sure must have a short attention span. Humans will likely be gone from the planet in a geological heart beat and our governments, nations, cultures and religions will become nothing more than a historical afterthought in an endless universe that has existed for nearly fourteen billion years."

"I'll tell you something I've learned while I dabbled with different religions. You might think they are all different but you know what? Nearly all religions are all basically the same in the end. Most of them, regardless of whether they worship one God or many, have four basic principles in common. Usually these four principles have been written down in books that are considered sacred like the Bible, the Qur'an or other holy texts."

Without allowing John to get a word in edgewise, Evan plowed ahead without a moment's pause.

"Most all religions have some kind of explanation for how the world came to be. Man has always wondered where all of creation came from and religions are not short on offering an explanation. However, all written religious explanations about creation are basically symbolic literature since none are based on any type of fundamental scientific understanding of how things actually work because they were written so long ago."

"Another thing common to most all religions is that they offer guidelines on how one is supposed to live. Rules if you will. The majority of these rules define how to treat one another as well as how to pay respects to the Almighty. Most religions tend to have lots of rules I'll tell you. Many of the rules come from man's interpretation of religious texts, but other rules are said to have come directly from God. Many of the rules are actually kind of silly really and don't make much sense if you stop to think about them. Nevertheless, religious rules do serve a useful purpose I suppose for keeping one's mind on the creator in a disciplined fashion."

"Most religions also have something to say about the afterlife, or what happens to a person when they die. Just as nearly every human with a half a brain wonders from time to time how creation came to be, most everyone would also like to know what will happen to them when they die. Not an insignificant question I'd say since dying is the one sure thing that every living creature is destined to experience. It's part of life really and dying is as natural as breathing. Can be a bit scary I guess if you have worries about what will happen to that little voice in your head after you croak. Anyway, nearly all religious books have something to say about the afterlife and usually these books have something to offer as well. Heaven, hell, reincarnation, nirvana, paradise…depending on what religion you follow you could end up in any one of a number of places upon your demise. Where you end up depends not only on which religion you follow but upon how well you follow the rules. The carrot and the stick the way I see it."

"Finally, in one fashion or another every religion speaks of the creator as a provider that is involved in everyday life. Sometimes this may go to the extreme and the creator is believed to have a detailed plan for everyone that is set from the moment of birth. This can be a mixture of good and bad things depending on the texts around which a particular religion is based. Good things God might provide include life, food, shelter, rain, sunshine, care, happiness, guidance, health, and wealth. On the other hand, examples of the bad things God might do to punish humanity or individuals are disease, plagues, natural disasters, pain, and general suffering. Always, however, the creator is depicted as a superior entity that is genuinely interested in human beings. The super-creator provides everything that humans need. God has plans for everyone, makes promises to them and at times punishes them when they break the rules. Apparently the Almighty even likes to hear from humans. At the very least the Almighty likes to hear a few words of thanks once in a while for what humans have been given. Some religions claim that anyone can have a personal relationship with the creator that involves daily, 24/7 access to the Almighty. A God who is on call, always ready and happy to listen to human thoughts and petitions directed through prayer, meditation and worship."

"Yep, there you have it. The central beliefs of all religions summed up in four simple words; creation, rules, afterlife, and provider…CRAP for short."

"Gee. For someone who has practiced so many religions you don't seem to think much of them. Don't you believe in God?" John asked looking kind of puzzled by Evan's seemingly cavalier attitude about religion.

"Oh, I didn't say that! I just don't believe in the God of any religion."

"Funny." Said John. "I got the idea that's kind of how Sam and Neva think, or don't think, about God. Actually…" John paused and thought for a minute, then shook his head. "You know, I really couldn't tell what they thought."

"Your not talking about my Sam are you? And who is Neva?" Evan asked with a curious and surprised look.

"Oh sorry, no." John answered. "Not your Sam. The Sam I meant is a boy I met earlier today and we talked for what seemed like hours about the creation of the universe. Then later I met Neva. She was a very nice lady. I talked with Neva for quite a while before coming here to your diner. Neva and I talked about the origin of life, biology and evolution. We talked about how long humans have been on earth and about evidence that different species of humans actually lived on earth together at the same time hundreds of thousands of years ago."

"Neither Sam or Neva seemed to think much about the Bible." John continued, grabbing the opportunity to interrupt Evan's sermonizing for a moment. "In fact, everything Sam, Neva and I talked about was so much different from the way the Bible talks about creation and man. I've no reason to think that Neva and Sam were just making this science stuff up. Besides, I've found it's fascinating to try and think about the distances, the vast periods of time and the complexity of life. How'd all this come to be? It's all just got me wondering why people believe so strongly in a God described in books.... especially when these books, if taken literally, are often in direct conflict with what science tells us. On the one hand, if God is all powerful and all knowing wouldn't a book like the Bible have correct information in it? I mean, if science is wrong about the age of the universe and evolution wouldn't the Bible at least be more accurate about what is correct? How can you believe literally what the Bible says about creation when it also says the earth is the center of the universe and that Noah fit two of every animal onto the ark? Did that include every species of beetles and was that before or after the dinosaurs became extinct?" John stopped and looked at Evan, his eyes searching for answers as if Evan were an all-knowing, all-comprehending fountain of wisdom and knowledge.

Instead, Evan just smiled. His kind blue eyes twinkled as he returned John's gaze and peered deep into John's eyes with a laser-like focus. John felt a warm sensation in his chest as his eyes met Evan's and John's inner voice, that constant internal chatter that is forever talking, became suddenly quite.

Evan spoke, "Son, you've got a lot of questions. Let me tell you what I think while you finish your pie. As you know, I've got an opinion on just about everything. Drives Sam crazy. For what its worth though, I'll give you my view of the world. Might take a while so I'll just come over and plop down on that stool beside you and we'll have a little chat if that's ok?"

Not waiting for a response from John, Evan yelled over his shoulder into the kitchen as he rounded the counter and took a seat beside John. "Sam, looks as if this is it for the night. Might as well turn off the neon's. I may be talking to this fella for a while."

Sam didn't respond.

John took another bite of pie. Delicious.

CHAPTER 27

The Bible Tells Me So?

Evan began. "Well, just because I've read most all of the holy books this doesn't make me an expert by any stretch of the imagination. But I do know a thing or two about most of them to understand that they all contain CRAP. At the same time, I don't want you to get the idea that CRAP is necessarily a bad thing. I guess ever since man took his first upright steps on this earth he's probably wondered why the earth, the sun, the stars, bugs and such are here. Anyone who can think has an inner voice that speaks to him. Human beings have a natural curiosity about what happens to them when they die. This is probably why several of those different human-like species buried their dead even hundreds of thousands of year ago and certainly long before any of the good books were around. I suppose it's part of what makes humans, human if you know what I mean. Human beings wonder about things and they are smart enough to construct possible explanations whenever no obvious answer to their questions exist. When things are beyond imagination, humans will fill in the void and make up stories about how things might be. In some respects this is comforting since if you can feel like you know the answer to a question, it can make the question less frightening or confounding even when the answer may just be human imagination at work. Think about the first time a human experienced a thunderstorm, fire, or a volcano erupting. Could be pretty scary I suspect. So what did man do? He created stories about things like why there is thunder and lightening, where fire came from, and why the earth erupts and pours out ash and lava. These stories involved the first gods, or more accurately, mysterious unknown and powerful entities that had control over things that man either did not understand or could not control himself. In the absence of any factual information, gods were created by man as an explanation for why things were the way they were, why there is pain, why things die."

"At some point humans began to record what they thought and how they viewed their world. Initially, early humans drew pictures on the inside of caves to record stories or memorialize important events. Later, as language began to develop in a more sophisticated fashion, a combination of both factual and mythical stories were passed along verbally from generation to generation. Eventually man developed written language and the dawn of the good books began. Over time, collections of writings became designated as 'holy' and these good books served as the focus around which different religions and ritualistic practices were based."

"Most of the good books were written a long time ago, at least by human standards. Mind you, not a long time ago in terms of creation, but a long time ago as far as recorded human history goes. Sounds like you already know that human history is not a very long time at all. Nevertheless, every one of the good books is supposedly either the 'inspired' word of the Almighty or, actual words from the Almighty that have been dutifully and accurately received and recorded by the creator's personally chosen human scribes many centuries ago. Every one of the good books contains a wealth of useful information to live by. Be kind to others, don't lie, take care the earth and the things you have. These are words that offer very good advice. However, good advice is not unique to any one religion and unfortunately, religions always seem to impose a lot of CRAP along with good advice. Unfortunately, the CRAP can become problematic if the good books are either interpreted literally or, as in many cases, are viewed as the divine and inviolable word of the Almighty."

"In truth, with a few exceptions, such as the Book of Mormon, most of these good books were originally not intended to be taken literally. The first written texts were considered 'mythos', meaning that they did not necessarily represent accurate historical events or factual information about anything. Rather, the early writers of holy texts were providing a descriptive guide for thoughts and ideas about events for which the true details could never be known or fully understood. Case in point....creation and the creator. In all likelihood, the early writers might find it amusing that their words are taken literally by so many today."

"Let me tell you something." Evan continued without hardly stopping for a breath. "Having read many of the good books I have to say that if these books are the creator's exact words dictated to humans and intended to be taken literally, the Almighty is a very chaotic thinker. In addition, if the creator choose specific human writers to tell 'the' story, God Almighty certainly didn't pick the best writers he could find...Mohammad, the Hand of Mormon, Mathew, Mark, Luke or John were not Shakespeare. If you pay attention to details, the holy books are filled with mistakes and inconsistencies. Seems that the creator of the entire universe would be a better proof reader don't ya think? Not only that, the good books are filled with hundreds upon hundreds of contradictions which make it very confusing to understand exactly what the Almighty is trying to say. For example, in the first chapter of Genesis in the Hebrew Bible or, old testament, it says that the Lord created animals and then created man. But in the second chapter of Genesis only one page later it says that God first made man, and then made animals. So which is it? The New Testament tells you to let your good deeds shine like a beacon before man, but then later it also says to not let anyone see your good deeds. And the Judeo-Christian Bible is not alone in its contradictions. The Qur'an speaks about wine and strong drink as Satan's handiwork but it also says that Paradise will flow with rivers of wine for the departed to enjoy. Come on now give me a break! Wouldn't you think that the Almighty would do a better job with the simple things in an instruction manual? But then on the other hand perhaps we need to give the Almighty some slack. Perhaps these contradictions and errors in the Almighty's words are like the human appendix…it seemed like a good idea at the time but it really turned out not to be necessary on second thought."

"One of the only published works that I'm aware of that the Almighty supposedly originally wrote by his own hand are the Ten Commandments. Apparently these were so important that the creator carved them onto stone, not once, but twice. I'll you tell though, I find the Ten Commandments a bit curious to say the least. To start with, the first four are all about making sure that humans show the author some respect. It doesn't sound like the creator of the universe has much self confidence if he was compelled to make the first 40 percent of the most important laws given to man all about how to pay God some respect. I just don't get the Ten Commandments. It seems to me that the designer of DNA, the creator of life, the hand that placed hundreds of billions of galaxies in the sky and created an infinite number of parallel universes could come up with something more profound than a rule about not taking his name in vain!"

"The remaining six commandments all provide some pretty good advice I guess. Honor your parents, don't steal, don't kill, don't lie, don't cheat, don't covet. The problem with these is that the Almighty plagiarized these last six commandments with reckless abandon. These commandments weren't the first time basic rules of living a good life were communicated to someone or written down. Even the godless, heathen Egyptians (that is according to the Israelites) had laws against theft, murder, lying and such long before Moses led the Israelites out of Egypt and was given the stone tablets containing the Ten Commandments. Wouldn't the creator of the entire universe in all its glory be able to improve on something that heathens had already thought of centuries earlier regarding how to behave in polite society? Even the golden rule, claimed by Christians as Christ's greatest teaching… to love others as yourself…. is not original with Christ. Many philosophers well before Christ's time including Confucius and Jina as well as ancient Greeks like Pittacus and Isocrates taught the golden rule more than twelve hundred years before Christ was supposedly even born."

"There are many other things in the good books that don't make any sense either. Both the old and new testaments, for example, are very fond of slavery. Most Christians today would agree that slavery is a bad thing. Although, I'm sure there are few folks here in Virginia who would like to still like to have their slaves back! Anyway, I digress. So back to slavery in the Bible. You have to realize that when these texts were written slavery was a way of life, deeply woven into the fabric of society. It was the norm. Therefore, its not surprising that the Bible has a view of slavery that is not consistent with how most of us think about slavery today."

"Even more confounding are the punishments and promises described in the holy texts. For instance, do you know what the penalty for breaking one of first six of the Ten Commandments is?" Evan looked at John expecting an answer.

John shook his head and said, "I guess an eye for eye?"

"Nope," Continued Evan. "Worse than that….death. There is nothing ambiguous about this. If you break a commandment, you are to be killed. Actually, if someone breaks the first two commandments, entire cities are to be destroyed. Now this seems rather severe I know, but that's what it is. The peculiar thing is, however, since the earliest establishment of every religion, believers have been picking and choosing which parts of the holy texts are to be taken literally and which parts are simply symbolic. Obviously this interpretation changes over time and depends on who is doing the interpretive filtering. However, I just wonder how anyone is to know which writings to take literally? What parts are indeed the precise word of the Almighty and which writings are to be dismissed as ancient symbolic literature?"

"It seems to me that if you are going to believe that some parts of a holy text are inviolable and 100 percent accurate, then you are obligated to believe that the entire text is the truth. Either these books contain God's word or they do not. There should be no room for a moderate interpretation of the literal word from the Almighty. In a way its not surprising to me at all that someone would fly a plane filled with innocent people into a building. The Qur'an contains language instructing true believers to kill infidels and it states clearly that if you are martyred in the process you will be rewarded in paradise. So, if you really truly believed deep within your heart and mind that your God wanted you to kill nonbelievers and that you and all of your family will be rewarded beyond imagination in the afterlife if you die committing such an act of martyrdom, why would you not do this? How is such an extreme belief any different from belief insisting that because the Bible says it is so, the earth can only be six thousand years old and the creator made everything the way it is in 148 hours? Both are supposedly the literal interpretation of the inerrant holy word of the Almighty and each is considered true by those who follow the literal word of those holy texts."

"Fundamentalist beliefs belonging to different religions catalyzes a great deal of the conflict between the Evangelical Christian, Orthodox Jewish and Muslim extremist worlds. However, in reality, any one of these beliefs is no less extreme than the other if members of each faith choose to interpret literally as true the words of their God as recorded in their respective holy texts."

Now on a roll, Evan became more animated as he continued speaking, gesturing for emphasis and shaking his head for emphasis as he spoke.

"It's impossible for me to believe that these so-called holy texts upon which today's religions are based reflect in any way the thoughts or actual words that have been transmitted to us by the Almighty. With few exceptions, these texts were written at least a thousand or more years ago. The oldest ones are from around 2500 BC and are filled with mythology and mystery. These books reflect man's natural thoughts and ideas about the creator as they try to address questions about why we are here and how the universe came into existence. It is not surprising at all that these books contain similarities since they were all written by man."

"This isn't to say that some of the writers did not think they were transcribing radio signals from above. There is no shortage of kooks in the world, either now or when these texts were first authored. Even the most pious believers today would be highly skeptical of anyone who says they receive personal messages from the Almighty, especially when those messages cause someone to do something that is evil, or at the very least, causes one to act in a way that we believe to be abnormal. But, you know what…read the Bible, the Qur'an, the book of Mormon…any of them. They are full of abnormal and weird things that supposedly the Almighty instructed people to do…. from the completely mundane like not wearing clothing containing mixed fibers, to the extreme such as the oppression of women or ordering someone to kill their own children."

"I could argue, as many have, that there is no such thing as a creator in the first place. If you believe this then none of the holy texts matter at all other than to represent forms of ancient symbolic literature. Atheism removes the intellectual burden of having to rationalize the realities of life with what is written in the holy books. That's all well and good so long as you are willing to concede that everything, all of creation, came about purely by random chance. Billions of years is a very long time and who's to say that things just couldn't happen over such long periods of time."

At this, John interrupted. "Yes! Neva and I talked about this very same thing. A billion years is a very long time. However, I just can't believe that even over billions of years, molecules just randomly collided to form DNA and then cells came about and started to multiply. Once you have a cell, or a bunch of them, I guess I can see how this could serve as the seed for evolution and over billions of years most anything could happen. But it's just too much of a stretch for me to believe that life itself just spontaneously began. And if you think about the Big Bang, you are faced with the question of where did the stuff for the Big Bang come from? I don't know, there are so many unknowns in creation that tell me there must be something responsible for the beginning…. at least I guess I'd like to think so."

John stopped and again looked hopefully at Evan for an answer. He'd not touched his pie since Evan had launched into his manifesto on the good books. John was completely spellbound by this man with the stubbly beard, resonant baritone voice and piercing blue eyes.

"Well I'll tell ya…" Evan began, then paused, scratched the stubble on his chin, took a deep breath and continued. "That Neva sounds like a real interesting woman I'll say. I'd like to meet her. One thing is for sure, nobody knows now do they? Sure, if you believe chapter, line and verse of one of the good books you may think you know exactly how all things came to be and why. But really, no one was there, right?"

Evan gave John a little wink and went on.

"You just need to ask yourself how much faith do you want to place in the precision of a so-called holy text that may have been written thousands of years ago. Then, you might need to pick and choose what to believe as fact, and what is fiction. It's not that recorded human history in some these good books is not important or that they don't have something important to say about morality, caring for others and the like. The issue is that these books simply have nothing accurate to say about creation, time, space, evolution, or any of the science we know today."

"Let me tell you a little story. I've something at home that means a great deal to me. It's called an ammonite. Ever heard of that?"

John shook his head.

"An ammonite is the fossil of an extinct creature that lived in the oceans about 400 million years ago. It looks like a spiral seashell. Mine is about four inches in diameter but I've seen them as big as a foot or more across and other's are as small as your thumbnail. My ammonite has been cut in half and beautifully polished so you can see all the grains of sand and silt that filled each chamber of the shell after the ammonite died and was turned into stone over eons of time. Looks like jewels on the inside. The ammonite is the very distant relative of modern day squids. In medieval times, that is up until about 1500 AD, ammonites were thought to be petrified snakes! Before the sciences of paleontology, geology and evolutionary biology uncovered the truth about ammonites the Catholic church proclaimed that ammonites were snakes that Saint Patrick had turned into stone as one of his many miracles qualifying him for sainthood. Needless to say that was wrong! But as far as Christians in the middle ages were concerned, ammonites were petrified snakes made by Saint Patrick. Makes you wonder what else the church might be wrong about doesn't it?"

"Anyway, that's neither here nor there. Sorry. My point is that when I hold this ammonite in my hand, I'm holding the remains of a creature that lived on earth hundreds of millions of years before even the dinosaurs existed! This creature had only a very small brain. It couldn't think. It didn't spend time admiring the beauty of its surroundings and it certainly never gave a thought to why it was here or how it would be remembered in history. Largely, the ammonite simply drifted with the ocean currents in an aimless and gentle dance until it died and slowly sunk to the bottom of a primitive sea. Once it arrived at the bottom of the ocean the dead ammonite took its place in layers upon layers of debris amidst the carcasses of other creatures that lived and died before and after its time. In this watery and silty tomb at the bottom of the sea, the ammonite would eventually turn to stone as it passed away the millennia. Entirely new species of life would come and go while the ammonite slept in its rocky tomb. Oceans deepened and receded. The earth would freeze and thaw. Giant asteroids would pound the earth wiping out hundreds of thousands, maybe even millions, of species in the aftermath of each collision. Mountains would burst skyward from the earth. Then, eroded by wind, ice and water, those mountains would be turned into valleys and deserts. The ammonite would watch as monkeys and man branched off in separate directions on the evolutionary tree of life."

"Finally after hundreds of millions of years, perhaps in Canada, the Sahara desert, or some other far corner of the earth that had once been a covered by a great ocean eons upon eons ago, a fossil hunter discovers a clump of ordinary rock. Perhaps just an edge of the ammonite was protruding from the surrounding rock bed where it had been entombed for nearly 400 million years. The fossil hunter carefully chiseled the ammonite out of its rocky tomb, freeing it from the surrounding shale so that the fossilized ammonite could be taken to a lab or workshop. The ammonite would be cleaned and cut in half. The inside would be polished until it shown like glass giving it the jewel-like appearance of a fine work of art. Eventually this particular ammonite winds up in some touristy junk shop where I find it. It speaks to me. It grabs me by the heart and says, 'I am immensely old. I've seen life come and go over eons of time. I was alive when no creatures walked on land.' Because the ammonite speaks to me, I buy it, probably paying far more than it is worth. But I like it and I take it home and place it on my mantle."

"The ammonite shape is really interesting ya know. It's a perfect exponential spiral. You've seen these often I'm sure. The shape is reproduced countless times in nature including: the spiral of a seashell, the arrangement of the seeds of a sunflower, eddies in a river, the pattern of nerves in the eye's cornea, a hawk's circular spiral as it searches for prey from above, hurricanes, the Milky Way galaxy and hypothetical portals to parallel universes. All of these things and many more examples found in creation are in the shape of a perfect exponential spiral. There are plenty of mathematical explanations for the natural preference for such a shape but the consistency of this repeated pattern from ammonites to galaxies hundreds of thousands of light years across is breathtaking."

"When I hold this little ammonite in my hand, I try to imagine what it has witnessed over hundreds of millions of years. I think about how all living things are connected through molecular form and function. At one time this ammonite contained DNA just like you do John. Some of the proteins the ammonite used to feed, build and heal itself are identical to the proteins in your cells today. The ammonite lived and reproduced offspring, perpetuating its species over millions of years. The ammonite is made of atoms that were once part of a star that formed and exploded billions of years before the sun and solar system were formed. This, my friend, is the story of creation and the continuing adventure of life. No holy text or religion on earth can teach us as much about the creator as my little ammonite. The Old Testament of the Judeo-Christian Bible may be three thousand years old but that is merely the blink of an eye to the ammonite."

John turned on his stool and stared at Evan. His day had started off like any other. He'd gotten up early, showered, shaved and dressed. A quick breakfast and then a mad dash to the first of several meetings. In the car driving home he was in control of his life, his mind, and his world. Then came the deserted rest stop. John's day evolved into a crash course in astrophysics, molecular and evolutionary biology. Now he was sitting alone in a diner talking to an old man about Biblical theology and fossilized sea squids. Despite the total weirdness of his day John was having a wonderful time. Everyone he'd met had been a complete joy to spend time with and he'd found the conversations fascinating, so much so that he did not want it to end. He thirsted for more.

"That's a beautiful story." John said looking at Evan with a nod. "You really love that ammonite don't you? After hearing you talk about it, I can see why. Its almost like the ammonite is your religion."

"Creation is my religion." Evan corrected. "I can find everything I need and the answers to my most profound questions by simply looking at creation and thinking…it is very good."

"Now, that sounds biblical." Said John as both he and Evan shared a laugh.

John glanced at his watch and realized that while it had seemed that he'd been at Sam's for a very long time it was not nearly as late as he'd thought. Although he felt an urge to get going, he had one more burning question on his mind. When he had first learned from Neva that during human evolution several species of hominids had co-inhabited earth together, John had begun to wondered whether or not other species of humans had souls. John was sure that if he asked, Evan would offer an answer. After all, Evan had an opinion about everything. The more he had listened to Evan and reflected about what Neva had talked about, however, John was now thinking that perhaps the right question to ask was not whether Neanderthals had a soul. Rather, it seemed that the correct question was more along the lines of what makes each of us human. Why are we curious about creation and its purpose? That little voice inside of our heads, our ability to love, to hate, to create, to wonder… is that inner voice what make us truly human and different from a chimpanzee or a Neanderthal? Neva had said that the Neanderthals and other species buried their dead so they must have had some beliefs about the afterlife. Alternatively, were Neanderthals and modern humans just branches on an endless bush that evolved higher cognitive functions causing our different species to think about themselves and the world in ways that animals cannot? Is man's "soul" simply the manifestation of a greater intellect and one of but many steps along an endless evolutionary path leading to who knows where?

John's thoughts began to drift toward wondering about what would happen to him when he died.

CHAPTER 28

The Phone Call

After reading a bedtime story to the girls as part of their nightly ritual, Lucy had put Naomi and Jessica to bed. She had just nodded off herself in front of the TV when the phone rang. Sleepily Lucy fumbled for the cordless phone she had had in her lap when she drifted off. Finding the phone she pushed the flashing talk button and said,

"Hello." Her voice raspy with sleep.

"Mrs. Mitchell?"

"Yes." Said Lucy, hovering drowsily somewhere between wakefulness and sleep. "This is she. Whose is this?"

"Mrs. Mitchell, I'm Dr. Sanger. I work in the emergency room at the Fredericksburg Memorial Hospital. Are you related to a John Mitchell who lives in Durham, North Carolina?"

"Yes." Lucy replied, foggy but suddenly worried. "I'm his wife. What's happened? Why are you calling me at this hour. Is John alright?"

"I'm sorry but I'm afraid your husband has been in an automobile accident. He arrived a short time ago at our emergency room and we are treating him now."

Lucy was immediately fully alert and she bolted upright on the sofa.

"Are you sure its John? What happened? Is he going to be all right? I need to get there. What am I going to do with the girls? Oh, please tell me he's okay?"

"Mrs. Mitchell, please try and stay calm. I know this is very difficult. Your husband is badly injured but he was conscious when he arrived at our emergency room. We are an excellent level one trauma center. Your husband is strong and he is in very capable hands. We have a great team here and are taking very good care of him. He's stabilized and we are assessing the extent of his injuries before we take him into surgery. If you can, you should try and get here as soon as possible since there might be some decisions that need to be made regarding treatment. Is there someone who could come with you? You shouldn't make the trip alone. It would be good to have someone with you for support." Dr. Sanger paused, waiting for Lucy to digest what he had just said.

Lucy's mind was racing. *"Oh my God."* She thought as she fumbled with what to ask next.

Lucy had always thought John as invincible. Athletic and strong with limitless energy and enthusiasm, John was unfailingly optimistic and never, ever showed any sign of weakness, that is, except around his girls.

"Oh dear God, the girls." Lucy thought and her heart sank. *"How can I even begin to tell them that their daddy has been hurt? Or worse yet, what if he dies and have to tell them they can't see him any again. I can't do it, I just can't."*

Then, knowing that John would simply not accept any negative thinking no matter what the situation, Lucy closed her eyes hard and pinched the bridge of her nose as she struggled to get hold of herself. Taking a deep breath Lucy tried to focus on the things over which she had immediate control. After a moment, Lucy spoke slowly in a quiet measured tone, enunciating every word with care and determination. "I understand Dr. Sanger. Yes, I have a friend who I'm sure will drive me. I'll be there as soon as I can. Let me give you my cell phone number. Is there a number I can call to reach you?"

That was about all Lucy could say before her voice cracked and she fought back tears once again, adding, "Please, please take care of John. He is everything his girls and I have. I'm trusting you to not let anything happen to him."

"We're doing everything we can Mrs. Mitchell. We are one of the best trauma centers in the country. John is a strong, healthy man and he's in excellent hands." Dr. Sanger repeated calmly and with compassion, trying to reassure Lucy despite the difficult news about her husband.

Lucy and Dr. Sanger exchanged phone numbers and Dr. Sanger gave Lucy directions for the quickest way to get to the hospital which was not far off the first Fredericksburg exit on I-95 north.

Lucy thanked Dr. Sanger and hung up. It took a great effort and several moments to composed herself once again. With trembling hands Lucy hit speed dial number two and called her best friend Pam. Pam and her husband Steve lived just down the street. They had one girl and the two families had been close since they had both moved into the neighborhood about the same time four years ago. Lucy knew that Pam or Steve would not hesitate to drive her to Fredericksburg while the other stayed home to look after the girls.

After several rings, Pam answered.

"Pam," Lucy said, "John's been in an accident and I need to get to the hospital in Fredericksburg."

Time stood still.

CHAPTER 29
He's a Soul Man

"Looks like you've still got things on your mind son." Said Evan as he and John sat side by side. "Go ahead, lay it on me man. I've got an opinion on everything you know!" Evan repeated with a knowing wink and a smile.

"It seems..." John began, "...that much of religion is built around making people feel more comfortable about dying. You know, promising that the afterlife will be better than life on earth, no pain, no disease and so on. I've not thought much about dying to tell you the truth. I guess most people my age don't if they are healthy. We kind of feel invincible and act as if life will go on forever. But, I know it won't. Everything dies. Neva told me that 99 percent of all the species that have ever lived on earth are extinct today so there has been a lot of death on this planet over billions of years. Even all the different species of humans are gone except for us. We'll probably go too some day replaced by some new and improved human species."

"It has just got me wondering." John continued. "Just suppose there is someplace like heaven where your soul goes after you die. Will there be Neanderthals there? Did they even have souls? Will I see my uncle Harry? What about hell? Is the creator really going to toss people into the eternal fire because they told a lie or had the misfortune of being born in the wrong part of the world and practiced the wrong religion during their brief time on earth?"

"I don't know if there are answers to any of this. Probably not. But everything I've learned today about the universe, creation, evolution and religion sure has me wondering. For sure, nothing fits with what I remember learning in Sunday School and church about creation, Jesus, heaven and hell. From what you've said no religion is going to have the answers. Perhaps instead of teaching me about Jesus and the Virgin Mary, the church should have been teaching me molecular biology and ammonites!"

"That's a lot to think about." Evan said in response to John's questions. "What do you think? Do you think you have a soul?"

John thought for a moment. "I'm not sure. I have this voice inside my head that is constantly talking. I guess it's what makes me, me. It makes me feel happy when I've done something to be proud of or when I see my kids do something amazing. For sure it tells me life is great when I'm sitting quietly with my arm around Lucy. At the same time, this voice can make me feel guilty when I do something wrong or stupid. It talks to me all the time. It never shuts up. My guess is that everyone has their own voice. I'm pretty sure that animals don't have a voice…at least, I'd guess this is the case. But, I don't know, is this inner voice my soul?"

John continued thinking out loud before Evan could say anything. "If I'm just one in a long line of human evolutionary experiments when did the voice evolve into existence? Since mine talks to me I suppose it would have had to be when language developed. If there was no language my voice would have no way of communicating with me other than perhaps through some grunts and snorts with little meaning. So, if the voice is my soul, then souls might not have existed before there was language? In fact, if the voice is my soul maybe I didn't have a soul until I could understand language. So do babies have a soul?"

"When I think about something like how beautiful a sunset can be, or how large a hole in the universe is, does that mean my soul is pushing me to think about creation and a creator? If I'm inspired to do something good for someone, like help them change a tire, it's my inner voice that calls to me and says, 'you should stop and help that person change their tire'. When I think through a problem at work I literally have a conversation in my head with myself, exploring options and deciding on a course of action. If I do something wrong and feel guilty, is that voice really my soul telling me I've sinned? When I die, what will happen to that inner voice? Does it go silent or does it go somewhere else?"

John stopped and looked at nothing in particular, lost in his own thoughts for a few moments. Then, John sat upright, nodded and said, "Yes, when I think about it, yes. I'd have to say that my inner voice is probably my soul. At the same time I've no idea where it comes from or why it's in me." John stopped and turned to Evan, again looking for answers to impossible questions.

Evan smiled.

"Gee, Evan sure does smile a lot." Commented John's inner voice.

Evan took a deep breath through lightly clenched teeth, nodding his head. "It seems to me that you've got a pretty good perspective on your inner self John. In most religions that talk about the soul or sometimes one's spirit, the soul is considered to be one's thoughts, personality and emotions. As you concluded, its what makes each person uniquely themselves. Atheists argue that it is pure biology that creates personalities and that the inner self is a reflection of both a person's DNA and influence from the environment in which they are raised. There is only a few percent difference in human DNA that makes *Homo sapiens* different from monkeys. Is it that small percentage difference in DNA wherein lies the inner voice? For certain, there are personality properties such as shyness or aggression that have been linked to certain genes but this is not uniquely human since animals can display shy or aggressive qualities as well."

"Religious texts talk about the soul or spirit as being created or given by the Almighty. It's interesting that you raise the issue of babies having a soul since there is quite a debate as you know between the pro-life and pro-choice groups that basically revolves around whether or not a fertilized egg, a single cell, is human or not. Pro-life advocates believe unequivocally that a fertilized human egg, even though it is only a single cell it is indeed human and possesses a soul. Pro-lifers argue that the fertilized egg has the potential to grow into a fully formed human being and therefore it should be afforded all the protections of a human life. Pro-choice supporters on the other hand, argue that any cell in the body has the potential to grow into an adult under the right conditions because every cell contains the same DNA as a fertilized egg. Pro-life argues in return that 'humanness' lies not in the DNA but in the soul, which, is placed into the fertilized human egg by the creator at the moment of conception even though a fertilized egg lacks a mind to hear and understand its inner voice. Makes me wonder, however, how do we know that animals don't also have a soul? Even a beetle's brain contains more cells and is larger than a human embryo. Certainly a beetle's brain has a greater capacity to process information than a fertilized human egg. Does a beetle have an inner voice that instructs it what to do? I have no idea, but personally I would not reject the possibility out of hand simply because I believe that a human embryo is more holy than an insect."

Evan continued, "Suppose for a moment that your inner voice is your soul, or at the very least, your inner voice is the means by which you are in touch with your soul. Perhaps the most profound question for humanity then is what happens to the inner voice or soul when a person dies. Do you think that an infant talks to itself? Does a newborn ever think about death? No, probably not at all. Just like an animal all a newborn infant cares about is that it has food and is comfortable. Eventually a child begins to associate food and comfort with caring and nurturing, even love perhaps. It's interesting you know that children, until they are around four or five years old don't really begin to understand death. True, they don't like to be hurt but they don't associate pain with death at that age. Neither, it appears, do animals associate pain with death. It is thought, or presumed, that animals don't have a concept or fear of death even though some animals like chimpanzees, elephants, whales and dolphins do appear to exhibit a type of mourning when another of their group dies. Nevertheless, until children develop reasonably good language skills they simply don't comprehend death in a way that allows them to understand that when something dies, it's gone from this world forever. Children's earliest fears about death are typically about whether their parents will die. However, it is not so much the fact that their parents would be dead that bothers children when they are young, it's the unknown of wondering who would take care of them if their parents were not around. Children might also wonder what happens to someone when they die but typically not in the morbid sense. Children are curious, but they spend little if any time concerned that they themselves might die."

"At an early age, children also don't have an understanding of right and wrong in the sense of morality, good and evil. Just like animals that can be trained, young children initially develop an understanding of what is acceptable behavior and what is not. However, they don't comprehend 'bad' and 'good' until language skills begin to develop. Does this mean that morality is tied to language? Perhaps so but, perhaps not. Even before children can speak they begin to develop a sense of fairness and can show empathy. Before the age of two, children begin to make judgments about consequences of their actions. As a result, a child's behavior starts to be directed toward intentionally doing good in order to gain a reward such as a parental smile of approval, or avoiding doing bad so as to not get scolded. However, its not until language skills become sophisticated enough to allow them to express themselves that children begin to show a clear sense of guilt when they do something wrong."

"I could see that in my girls as they grew." John said, stopping Evan for a moment. "Before they could talk, they definitely learned what 'NO' meant. But as I remember, when Lucy or I would tell them 'NO', it didn't really seem to bother them. In fact, it would become somewhat like a game at times. The girls would test us to see if we would say 'NO' to something they would try to do. Then they would stop, turn and smile at us as if to say 'really!' Kind of funny how this would happen but now that they are older the girls definitely know what it means with Lucy or I say 'NO'. I can tell you for sure the girls are not always happy with that."

"Its really interesting to watch kids develop." Said Evan. "Every day is a new experience for them and their parents. What a challenge and a gift to raise a child. How old are your girls John?"

"The older one, Naomi is seven going on eight. Jessica is five going on forty." John said with a chuckle.

Evan smiled and continued. "Those are interesting ages for sure. I'll tell you why. At the same time that children begin to develop more sophisticated language skills, children in most cultures also begin to develop strong beliefs about metaphysical beings such as Santa Claus, the Easter Bunny, goblins, ghosts, the Tooth Fairy, the devil and the Almighty. Early on a child cannot differentiate between any of these entities, for in a child's mind, they are all very, very real. Santa brings presents. The Tooth Fairy takes my tooth and leaves money under my pillow. There is a boogie man under my bed. God listens to my prayers at night and takes care of me. To a child, these are all one in the same.

"Eventually, as language skills develop further and rational thinking about the physical world begins to take hold, children learn to question the logic of certain beliefs. Take Santa Claus, for example. Eventually children question whether it would be actually be possible for someone to fly around the earth in a sleigh, slide down chimneys and leave toys for every boy and girl on earth in one night. At first, this is not a problem because a young child only cares that they get toys. They could care less if Santa visits anyone else. As they mature and become less self centered and more knowledgeable about the real world, however, children begin to understand that other children get toys as well. They also begin to understand that some children are poor and, as a result, may not get toys. How could a loving Santa forget to bring toys to poor children?

"Your older girl is probably getting to the age in school she might hear some talk from older kids that Santa Claus is not real. Belief in Santa and other imaginary beings persists only as long as children think that magic is real, for magic remains the only plausible explanation for how Santa could visit every house in the world in one night and the fat little elf slide down a narrow chimney. Eventually, however, children learn that these metaphysical beings are not real. One by one, belief in Santa and other magical entities are exposed for what they are, merely mythical stories for the young mind. Children learn that these myths are perpetuated for a time by adults for children so that the young can experience a sense of wonder and mystery. It is only after a child's mind begins to reach a certain stage of maturity that the truth about Santa is revealed. The exception of course to this is the Lord God Almighty and, in some religions, the devil as well. The Almighty is kept alive in youngsters' minds through adherence to religious customs, reading of the holy texts, and most importantly the behavior of adults whom children observe worshiping and praying."

Evan continued, "Thousands of years before Christianity, Judaism and Islam, God or 'the gods', were supernatural forces and not always actual beings that were superior to humans. Rather, gods were seen as mystical and unfathomable forces that were not accessible to the human imagination and which could never be defined or understood. Very importantly, belief in these early gods required a high degree of ritual. Regular practice of ritualistic behavior and contemplative thought about the unknown were recognized as essential activities for anyone who wished to maintain an adherence to any kind of belief."

"I suspect that if children watched their parents and other adults writing letters to Santa every year and sitting on Saint Nick's lap at the mall every Christmas, a child's belief in Santa would likely be sustained well past the time that reason and logic argued against the reality of Santa Claus. This is where the concept of cognitive dissonance comes into the picture."

"Cognitive what?" John interrupted. "That's a mouthful of a word."

"Cognitive dissonance." Evan repeated. "It's not as complicated as it sounds." he continued with a smile, his white teeth gleaming.

"Simply put, it means that if you consistently behave in a certain manner, your beliefs will eventually conform to your behavior. Perhaps this is why religious customs and rituals are so important in the establishment of faith.

Remember the Sabbath and keep it holy. Pray five times each day facing east. Fast regularly. Read the holy texts unceasingly."

"Revelation about imaginary beings is occurring just as a child's inner voice matures and more sophisticated language and thinking skills develop. As a result, older children begin to ask deeper and deeper questions about their world, themselves, the complexities and consequences of right and wrong. Along with more sophisticated language skills children begin to listen more actively to their inner voice, or as you have concluded, their soul."

"It is around the time when children begin to become aware of self and develop a more mature understanding of right and wrong that in most religious cultures, a significant transition occurs. A belief in the Almighty becomes prominently reinforced during this formative time through ceremonies and rituals designed to recognize and celebrate a child's transition from childhood into adulthood as a full member of society and the religious culture into which they were born. Belief in an immortal soul becomes linked to a personal creator who listens to an individual's inner voice and therefore knows one's most intimate thoughts. The creator takes a personal interest not only in what an individual does, but what they think, good and bad. The Almighty cares for each person, wants them to be good and likes to hear from individuals through prayer and worship. It is at this time when a child begins to understand sin and mortality that religion becomes established as the gateway, through death, to something better for those who believe and follow the rules. Just when a child is developing an adult understanding of his own mortality and the finality of death, the one imaginary being remaining from their childhood provides a reason to not fear death as the great unknown. The Almighty offers comfort, protection and salvation to the faithful, righteous and obedient. Death becomes merely a portal through which one passes and, waiting on the other side, is God to welcome you saying, 'Hello John. I'm glad you are here'. In most religions death is no longer an end to be feared but a great beginning or transition to a better existence. A step toward the Almighty to be embraced and perhaps even desired."

"Even for people who come to religion later in life through study or personal crisis, comfort can be found in religious dogma through the hope for a better life both here on earth and after death. Always, this involves one's inner voice as eventually the inner voice proclaims that religion shows the path and faith is the vehicle to salvation. Could this happen without language? Who is to say since without language it would be impossible to assess. What is clear, however, is that language, coupled with recognition of the inevitability and finality of death is critical to the development of religious beliefs. If the inner voice is the soul, then it is indeed an understanding of one's mortality that calls an individual either to accept or to reject religious beliefs. On the other hand, if the soul is something different than the inner voice, then acceptance or rejection of belief would seem to be the result of something else entirely over which an individual has no control."

"In the end, only the dead know what happens to their inner voice after death… that is assuming that there is something after death to be known. Whether or not you choose to believe what any particular religion has to say about death is a personal choice. There are lots of options. Not all can be right and in fact, there is no way to prove that any are right or wrong. Having studied and practiced most religions it is clear to me that all suffer from the same dilemma. Each religion may claim that its holy texts are the word of the Almighty and each turns to those texts in order to explain life and what happens after death. For as long as man has walked the earth and buried their dead, man has had this question. Unfortunately, no religion can offer anything more substantial than the writings of man as an explanation for what the afterlife may be like. Religious teachings and rituals are all designed to guide man's inner voice and all religions proclaim to offer the true path to enlightenment and a relationship with the creator. Unfortunately, every religion falls woefully short of the truth because all religions are a product of man's inner voice."

Evan paused and wearily shook his head. Then without a second thought as though they were the best of friends Evan raised a thin wiry arm and placed his hand on John's shoulder. John did not move as Evan slid his hand down John's back slightly and gave him a friendly pat between his broad shoulders. Then, as natural as could be, Evan draped his entire arm across John's shoulders and gave him a warm, comforting squeeze. Evan was surprisingly strong and his firm show of affection got John's attention. He turned slightly to face Evan. This time it was John that looked deep into Evan's eyes.

Evan returned John's piercing gaze and said in a low, quiet voice that only he and John could hear, "The creator's message to man is written in the language of creation John. All the majesty of the universe and all the mysteries of life are available to anyone who chooses to look. Creation calmly and without pretense or malice cries out that death is simply a natural part of life. On the topic of a soul, however, creation is silent."

Evan gave John another reassuring squeeze, patted John's back once again and removed his arm. Minutes passed as he and John sat in the quiet of the now empty, peaceful diner.

After some time, John spoke. "Evan, this has been a wonderful conversation. I can't say that everything is clear to me but you have helped me think about religion in a much different way. It seems to me that if I want to learn more about the creator, the best thing I can do is to learn more about creation. As odd as today has been it's made me think about things for which I've previously never given a moment's thought. Because of today, the universe, creation, evolution, and faith are important to me now. I want to understand more and I want my girls to understand as well. I don't know where all this will lead, but today has started me on a quest to learn more about creation and the creator. I might even buy an ammonite!"

Evan smiled kindly and turned again to face John, his shining blue eyes holding John's gaze. "Enjoy your journey John." Evan got up, gave John another friendly pat on the back before he disappeared into the kitchen without another word.

John sat and stared at his half-eaten cherry pie. He reached for his coffee and took a sip. It was still hot. Surprisingly his ice cream had not melted into a milky mess.

"*That's strange.*" John thought to himself. It seemed that he and Evan had been talking for hours and yet his coffee was a hot as if it had just been poured. He looked at his watch and was shocked to see that only a few minutes had passed since he had walked into the diner and ordered a cheeseburger. "*Couldn't be.*" John thought. "*It must have been earlier than I thought when I came in.*" He surmised.

"You haven't eaten much of your pie."

John looked up and saw Sam standing in front of him, her gaze shifting from the pie to John's face.

"Evan said you had a good chat. Can I get you anything else?"

"No, no thanks." John said looking at his watch. "Just my check. I need to get going. I can't wait to get home and see my wife to tell her about the day I've had. She'll never believe it."

Sam handed John his check as he finished the last few bites of pie and drank the last of the hot coffee. He was wide awake, refreshed and ready for the rest of the drive. John glanced at the bill and immediately noticed that, true to Sam's word, the pie had been on the house. John swung around on his stool, stood up and took out his wallet. He laid a twenty dollar bill on the counter and told Sam to keep the change.

"Thanks Sam, for everything." John said. Then straining to look back into the kitchen, John said, even though he could not see anyone, "Evan, thanks. I appreciate your time. Everything was great."

No answer.

The parking lot was now deserted as John left the diner and got into his car. He started the engine, backed up and turned to leave. As he stopped and shifted the car into drive, John noticed Sam leaning in and reaching toward one the windows to lower the venetian blinds beside each one of the booths that surrounded the interior perimeter of the diner. Catching Sam's eye, John waved as she lowered the last blind. Although it was dark and the distance made it difficult to see detail, John would have sworn he saw Sam give him a wink as he rolled out of the parking lot toward the road. Sam turned toward the kitchen and the lights inside the diner went out. John stopped for a moment and stared again toward the diner. He thought is was his imagination but it looked as if just before the lights went out and Sam turned back to the kitchen, she bore a remarkable resemblance to Evan. "Well..." John chuckled to himself and his inner voice said, "...*Sam did look like Evan in drag. I wonder if in some parallel universe, Sam is the cook and Evan is the waiter.*"

John stopped at the BP station he had noticed on his way to the Wendy's, filled up his tank with gas and visited the rest room before getting back on I-95 south. Accelerating on the entrance ramp to the highway, John merged with traffic on I-95. Once at a comfortable speed, John dialed home on his cell phone.

CHAPTER 30
I'm On My Way Home

No one knows how, why or even if it actually happens. Nevertheless the laws of physics are compatible with the existence of an infinite number of parallel universes. A cautious foot step onto a frozen pond causes a single crack to appear on thin ice. The solitary crack creeps slowly, yet relentlessly across the surface of the ice until without warning it encounters some unforeseen event along its path resulting in the birth of a new fracture line as one path becomes two. Each new crack continues their separate, but linked journeys until they too fracture. Eventually, the entire ice surface is a spiderweb of hairline cracks all linked and born from a single initial event. Might that be how the universe works? A tiny insignificant occurrence causes the universe to suddenly split in two, duplicating time and space across infinity. Each new universe identical in every way to the original, sharing a common past but different futures.

John's thoughts meandered between giant holes in the universe and how many Little Debbie cakes could be in the truck in front of him when the green Audi appeared out of nowhere. Had he been day dreaming? John didn't have time to consider where his mind had been. It all happened so fast.

At seventy miles per hour, one hundred and three feet per second, a mere fraction of a second can mean the difference between a near miss or catastrophe, life or death. Just one second more, or less, to adjust a seat belt or check the mirrors. One second more, or less, to smile at the thought of recalling the sound of your children's laughter. One second more, or less, to admire the sunset. One second to linger and enjoy, or reluctantly pull away from the warmth of a hug, a tender kiss. The thing is, one never knows before hand whether one second more or one second less would make the difference. One second to remind us just how fragile life can be, as it has been, for nearly four billion years.

The Audi dove in front of John. He slammed on his brakes and grimaced as he prepared for impact. The Audi missed his front bumper by only a few feet. In a flash the Audi was past the truck.

"What a fucking idiot!" John screamed out loud. "He's going to kill someone. Jeeesussss that was close."

Still a bit shaken from his near miss with the Audi only moments earlier, John noticed a rest stop ahead and decided it would be a good idea to pull off to stretch his legs, take a leak and get something to drink. Like any typical rest stop along a major highway, people were milling around going to or from the restrooms, walking dogs, or just having a short break before getting back on the highway. After visiting the rest room and getting a coke from a vending machine John lingered a few minutes to just look around at the people. All shapes, sizes, colors, ages and manner of dress were represented. Walking casually back to his car, John noticed for the first time the beautiful sky. The sun was beginning to set and the vivid array of colors painted across the western sky was spectacular. With his attention turned toward the sunset John was not watching where he was going. Suddenly, John's personal space was invaded by a teenage kid wearing baggy pants, a sweatshirt and a baseball cap. The boy was walking in the direction of the restrooms. With his headphones on, head down and totally oblivious to anything or anyone around him, the kid was just as distracted as was John and neither were looking where they were going. The boy and John stopped abruptly and contorted their bodies clumsily as they tried to avoid running over one another.

"Sorry mister." The kid said looking up briefly. The boy immediately continued his march toward the bathroom, head down once again.

"No problem." John responded, thinking to himself, *"Geeez I hate it when someone wears their baseball hat backwards. Makes them look like an idiot."*

After reaching his car, John paused for a long moment to soak up more of the setting sun. The brilliant colors were beautiful. John just stood and stared for a few minutes thinking to himself, *"That sunset is really going to be beautiful in a while. This world is really amazing."*

John got back on the highway and immediately called home. After a couple rings, Lucy answered.

"Hello."

"Hi honey," John said. "Good news. Things finished up early in DC and I'll be home earlier than expected."

"That's great!" Said Lucy, obviously happy with the news. "Do you think you'll be home before the girls go to bed?"

"Not sure, but I'll try. I'm only about an hour from Richmond so I should be home in about three hours or so depending on traffic. Tell the girls I should be home soon. Since tomorrow is Saturday, keep them up as long as you can, OK? Can't wait to see all of you."

"We'll be glad to have you home honey. Don't rush, you know how I worry about you driving, especially after dark." Lucy sounded worried.

"Don't worry. I'll be careful. Do we have any plans this weekend?" John said.

"Nothing really," Said Lucy. "Why?"

"Oh no particular reason. I just thought we could go to a book store maybe. I heard this report on NPR about a hole in the universe that is something like a billion light years across! The report also said something about parallel universes and that there may be other universes that have the same hole it them. Is that crazy or what? Anyway, its got me thinking that I'd like to learn a little about astrophysics or whatever it is that studies holes in the universe to try and get my head around how big a billion light years is."

"Sure, what ever you want." Lucy said with a laugh. "The girls always like to go to the book store. Drive carefully and please don't rush. See you when you get home. I love you."

"Love you too sweetie. See you in a few hours. I'll be very glad to get home."

John hung up and focused on the road while the voice in his head tried to imagine once again a hole in the universe one billion light years across.

CHAPTER 31
Parallel Paths

Lucy flipped her cell phone closed. She had just spoken again with Dr. Sanger to get a status update on John and to let Dr. Sanger know her estimated time of arrival. Dr. Sanger had given Lucy an update on John's injuries and told her that John would very likely still be in surgery when she arrived at the hospital. When she arrived, Lucy was to go into the emergency room entrance and ask the nurse at the admittance desk to page Dr. Sanger. He would meet her and take her to the surgical waiting area where they could talk.

"I just know John will be fine. He has to be." Lucy said quietly.

Pam nodded in silent agreement, giving a supportive smile as she reached over and squeezed Lucy's hand. "I'm sure they are taking very good care of him," She added.

"Please let John be alright." Lucy offered a silent pray as she and Pam drove north toward Fredericksburg, VA.

Pam and Lucy had talked only sporadically for nearly an hour since they left Lucy's home and headed north. Mostly Pam listened, offering words of encouragement now and again as their conversation roamed back and forth, peppered by periods of silence. Lucy talked about John, the extent of his injuries and the accident. She wondered aloud what specifically had happened and why. From the description she had gotten from the doctor, Lucy knew that it was bad and that John had been seriously injured. Pam was comforting and supportive but she also knew that there was little she could do right now other than to listen if Lucy wanted to talk, and remain quiet if Lucy seemed lost in thought.

How Do the Fools Survive by the Doobie Brothers had just begun playing on the car radio. With the volume turned down low, neither Pam nor Lucy had even noticed that the radio was on as they talked.

"Are you listening to the radio? I could turn it up if you'd like the distraction." Pam asked, breaking the silence in the car.

"Excuse me, sorry." Lucy responded. "What did you say?"

"Would you like to listen to the radio?" Pam repeated

"No thanks." said Lucy adding, "I just don't understand why this could have happened."

Pam turned the radio off.

Epilogue

A short time after talking with Lucy, John became bored thinking about holes in the universe and turned on the radio once again. Finding an oldies station out of Richmond John turned up the volume to help keep himself awake.

The pounding piano introduction to *How Do The Fools Survive* by the Doobie Brothers began to rumble from the speakers of his BMW and John tapped his fingers on the steering wheel along with the beat. The song's lyrics tell of a creator who, long ago, gave life to humanity and then left man on his own while a background chorus of human voices cries out repeatedly for God to show them the way and the light.

As Michael McDonald, lead singer of the Doobie's began to sing in his distinctively throaty voice, John roared passed the exact point in space and time where Lucy and Pam had just been. John traveling south. Lucy heading north. As they pass on opposite sides of the highway on this March evening neither is aware of the other for they travel along different, but the same, distant roads.

The creator smiles as new life blossoms. Sprouting from an existing branch, a new bud forms on the tree of endless creation. A new universe is born. But for one tiny mutation, the new universe is a perfect copy of the original including a hole in space a paltry one billion light years across. Another dimension, another creation, perhaps only inches or maybe billions of light years away from an endless number of other parallel worlds where creatures exist without the faintest knowledge of their infinite number of identical twins. Like all new things, creation has tinkered once again with life and another small experiment has begun. A new path, a new beginning. The creator beams with pride. And it was very good.

Concluding Thoughts from the Author

One of my favorite songs is by the Doobie Brothers called *How Do The Fools Survive*. Due to its length at nearly five and a half minutes, which limited its air-time on pop radio, it is one of the Doobie's lesser known tracks from their *Minute by Minute* album released in 1978. In addition to having one of the all time great guitar riffs at the end, the lyrics have spoken to me since I first heard the tune. The song was written by the group's lead singer and Grammy award winner Michael McDonald in collaboration with Academy and Grammy award winning lyricist and song writer Carol Bayer Sager. The lyrics are the haunting reflections of a perplexed creator who is persistently bombarded by a vapid chorus of prayers from humanity. The creator declares that he gave man life so very long ago and then left humanity on its own to make its own light and find its own way. The creator laments that while humanity has more stars than it has wishes, man foolishly gives up its responsibility and repeatedly asks God to show them the light and the way. The creator shakes his head and wonders how the fools survive.

Most of the world's major religions claim one or more ancient texts or sacred writings as the inspired word of their God. Based upon these holy texts, for better or for worse, organized religions have built world-wide faith-based communities of followers who, to varying degrees of extremism, live their daily lives according to the interpretation of the written words contained in these texts.

When people's knowledge about even the most basic scientific principles of biology, chemistry and physics is lacking it becomes impossible for the citizens of a society to understand and effectively interpret factual information about how the world in which we live operates. Often this basic scientific ignorance, when coupled with a strict interpretation of biblical teaching, creates a global ideological battlefront whereby biblical stories are elevated to the factual equivalency of modern scientific understanding. In but one example of this conflict between science and religion, some Christian fundamentalists and evangelicals reject the theory of evolution and have successfully pushed the ideological mandate for teaching biblical creationism and intelligent design along side evolution in american schools. This type of misguided religious fundamentalism clouds human judgment and erodes natural human curiosity about the world around us.

There are consequences for a failure to distinguish myth and superstition from fact. The wholesale and unchallenged acceptance of holy texts as the direct word of God and the strict interpretation of those texts results in religious extremism in all forms. This is important because based on religious beliefs, people fly airplanes into buildings and murder thousands of innocent men, women and children. This is important because based upon ideologies gleaned from literature written thousands of years ago by authors with zero factual information about the universe and creation, governments make decisions and enact sweeping policies on critical social and economic issues. Biblically influenced governmental policies impact laws that regulate things such as stem cell research, abortion rights, the treatment and prevention of AIDS, who can marry whom, basic human rights, climate change, and when to go to war. These life-changing policies are supported, and in many cases demanded, by a citizenry lacking even the most rudimentary knowledge of the science behind the issues. This is important because after three police officers were brutally gunned down in a Pittsburgh suburb a local priest believed it is comforting to tell his flock and the families of three murdered police officers that these men died because God needed them to come home for Holy Week.

A number of self-described atheists including Richard Dawkins, Daniel Dennett and Sam Harris have written extensively on the topics of evolution and religion. They use sophisticated scientific knowledge, logic and an understanding of biblical literature to present lucid and compelling arguments against the existence of a God. They attack equally as effectively so-called "intelligent design" which, is the modern day equivalent of watered down creationism. These atheist authors point to the futility of prayer and the failure of religion to bring about peace and harmony on earth as evidence that God does not exist. Dawkins, Dennett, Harris and others point out that nearly all of the world's conflicts since the dawn of recorded history, as well as the most egregious and barbaric examples of man's inhumanity to man have at their core direct religious motivation, zealous religious intolerance, and blind religious beliefs. Religiously centric and faith-based atrocities cited by these authors include the Inquisition, the Holocaust, the Salem Witch Trials, the events of 911 and the bombings of abortion clinics to name but a few.

Acknowledging that religion may have some personal benefits through encouragement of practices of meditation and personal reflection, Hawkins, Dennett and Harris argue that even moderate religious beliefs and tolerance are ruinous to society. Their reasoning is that the acceptance of religious beliefs and teachings as written in biblical texts is simply a matter of degree and that the most strict adherence to religious teachings and the written word of God ultimately and inevitably result in religious extremism in various forms. That is to say that moderate religious belief must, by definition, condone and even admire strict religious belief. The rationale behind this supposition is that while a moderate religious believer may select which parts of the biblical texts to follow, the religious extremist finds no room for biblical interpretation and they, the fundamentalist, in fact are simply more religious than the moderate believer.

Unfortunately, authors such as Dawkins, Dennett and Harris typically do not differentiate between religion and belief. Religion is defined most broadly as the organized practice of a set of beliefs. Belief on the other hand, can be sub-divided as either: a) the founded (testable), or b) the unfounded (untestable acceptance, i.e. faith) understanding that something is as it seems.

An example of a founded, or testable belief, is that the sun will rise in the east tomorrow. You can believe this is true because the sun has risen in the east every morning for as long as recorded history has existed. The rising and setting of the sun are governed by well understood and predictable laws of gravity, planetary movement and physics. The validity of such a belief can be easily tested by keeping a daily record (data) of every time the sun rises in the east. While you can never be 100 percent certain that the sun will rise tomorrow in the east, past history and the data collected would allow you to be quite confident in a belief that the sun will rise in the east tomorrow morning.

In sharp contrast to a belief that the sun will rise in the east tomorrow, stands the belief by Christians that through baptism they are saved and will go heaven to live with Jesus when they die. Islamic faith too teaches that if one dies as a martyr they will be immediately taken to paradise to enjoy the pleasures of seventy virgins. Unlike a belief in the sun rising tomorrow in the east, however, there is no way to test these religious beliefs in an afterlife. Because there is no valid record of anyone ever having experienced such occurrences and it is impossible to provide data that either supports or refutes a belief in the afterlife, such beliefs must be based solely upon faith. In this sense, faith is the unsupported acceptance of something that can neither be documented or experienced.

Atheists frequently employ science, logic and robust data supporting the process of evolution to argue against the existence of a God. However, what Dawkins, Dennett, Harris and others fail to distinguish effectively is the difference between the evolution of life and the origin of life. The origin of life should not be confused with intelligent design, or creationism in disguise. While intelligent design allows for some minor evolutionary tinkering with creation (that is for everything but man), it concludes that the tremendous complexity and vast diversity of living creatures argues for their specific and intentional construction by a powerful and thoughtful creator. The example most often used is that if you happened to find a watch on the ground you would naturally conclude that it was built by someone since it is far to complex to have essentially self-assembled out of nothing. However relevant and valid this argument may seem on the surface, it represents an illusionary misdirection and slight of hand by creationists. Remember when John and Neva were talking about the requirements for life, Neva taught John that the two basic elements for life are the abilities to: 1) sustain a source of usable energy, and 2) reproduce. A watch can do neither of these.

Certainly, as atheists would argue, it is true based on purely mathematical probabilities that the random collision of atoms over eons of time could have resulted in the formation of some organic molecules. Probability calculations can generate the mathematical odds for such an occurrence and those odds, while small, are not infinite. Nevertheless, even a rudimentary understanding of the basic workings of the most simple metabolic pathways coupled with the knowledge that sustainable life must be able to produce energy and to reproduce raises the probability for the random origination of life to beyond infinity. Once one understands some simple biology, the random origin of life would seem to become essentially improbable. The evolution of life over time due to random mutation and selective pressure is a scientific certainty. On the other hand, the odds of life itself originating purely by random chance are far, far less than your winning every state and national lottery in the world on the same day using the same number combination while you simultaneously hit the million dollar jackpot on a slot machine in Las Vegas at the very moment that a long-lost wealthy uncle you never met dies and leaves you his entire fortune.

Author's such as Francis Collins *(The Language of God)* and Gerald Schroeder *(The Science of God)* use science to argue for the existence of a God. In doing so, they launch from the basic premise that the biblical texts underpinning most of today's religions have some basis for validity. They argue that cold scientific principles and faith in a loving and benevolent God are not mutually exclusive. Indeed, both authors point to clear examples where biblical texts offer scientific insights which they claim are precisely in line with our technical understanding of creation and the evolution of life on earth. The problem with these approaches is that they begin with faith in God as the base and build from there to establish (or support) beliefs that are in agreement with the basic tenants of faith in a loving and benevolent creator. Unfortunately in science, one must begin with the data and move from the data to establish a set of conclusions or theories that are consistent with observations that can be further tested. Belief that the creator is good and cares for his creation is based solely upon faith and can not be tested in any fashion. Despite the surprising and awe-inspiring elegance embodied in a molecule of DNA, this does not prove that its creator must be good and loves human beings.

A key difference between beliefs based in science and beliefs based upon faith is in the believer's willingness and ability to adapt and change their beliefs over time. Scientific understanding is always changing. One only needs to read books like *In the Footsteps of Eve* by Lee Berger or *Evolution: What the Fossils Say and Why it Matters* by Donald Prothero to see how our scientific beliefs about the evolution of man and animals have changed dramatically over just the past fifty years as new data has emerged. While the fundamental process of evolution has not changed, our understanding of the process, the timelines over which change occurs, and the identification of key branch points on the evolutionary tree have been altered significantly as new science becomes known. Old ideas have been thrown out and have been replaced by new concepts based upon a deeper understanding of the basic principles involved as more robust data either supports or refutes earlier speculation.

In contrast, dogmatic religious beliefs may only grudgingly change over centuries in the face of overwhelming truths that unequivocally dispel antiquated faith-based beliefs about our world. Changes in faith-based beliefs, such as the belief that the earth is the center of the universe, are only officially acknowledged by religions after the world's population at large has known for centuries that the original faith-based beliefs were in error. Glacial-like movement in ideological thinking by organized religion only occurs when formal acceptance by the church can no longer cause a tear in the fundamental fabric of faith.

As I have studied the evolution of man and the wonder of all creation, I am completely and unashamedly humbled in a most reverent manner. Since a hominid first walked upright four million years ago on earth and the bushy branch of human evolution began, the path that man has traveled is peppered with failed experiments, fits and starts. The arrogance of belief systems that place modern man in a special place at the pinnacle of all creation is naive beyond reason and understanding. A vast expanse of time has seen numerous and varied sub-species of man that have walked, thought, explored and loved. Throughout countless millennia life has evolved and this life just happens to also include man. Anyone with an open mind can readily see and understand that our species' time on the tiny planet earth will surely end in just the blink of an geological eye.

What modern *Homo sapiens* share with each other and all living things goes back to the dawn of creation nearly fourteen billion years ago when the first atoms were created during the first few seconds following the Big Bang. The DNA, RNA, proteins, ATP, metabolic pathways, mitochondria, cell membranes and nuclei that comprise each of the one trillion cells that make up our bodies are not fundamentally different from those of any of the many trillions of creatures that have come an gone from the face of the planet since life first arose nearly four billion years ago. Indeed, the structural and molecular ties that bind humans together with all other forms of life that have ever lived on earth are far greater, more fundamental, and more faithful than any imaginary and mystical special tie we may claim to have with our so-called "creator".

You and everything around you including the pages in this book and the ink you are reading as words are built from stardust that is nearly fourteen billion years old. The oxygen molecules you inhale as you read this sentence are the very same oxygen molecules once inhaled by a dinosaur more than 250 million years ago. That same oxygen was also inhaled by your distant hairy relatives and Neanderthal neighbors ten's of thousands of years ago. These are the ties that bind us together; they are fundamental; they are ancient; they are real. Common links such as molecular and cellular structure do not require man made ideology for validation. One can only marvel at the timeless improbability of life and stand in awe of creation's complexity and yet, its beautiful simplicity. An unimaginably vast universe filled with untold opportunities for life is a testament to a creator that cannot be prescribed or defined by man's feeble and arrogant attempts to elevate ourselves as a favored part of creation. Such efforts make a mockery of creation and the nameless forces responsible.

Until humans realize that we are part of a continuing, self-perpetuating experiment grounded in biology, chemistry and physics, we have little hope of ever learning to respect one another or live together in peaceful harmony. Without some basic understanding of science, neither can we hope to understand that the planet we inhabit is the most improbable of all gifts. Modern religions offer no safe harbor for the evolution of a species. While a creator may have fashioned everything which is known including the galaxies, stars, planets, and the atoms comprising them, it is man and man alone that created religion in an attempt to explain that which is unknown. In this sense, a species with the ability to wonder and reason will always seek to explain what it cannot reasonably understand. However, what is crucial to a species' successful evolution throughout the millennia is its ability to adapt to change. The keys to longevity for any species are to accept and interpret new information, then translate this knowledge through the invocation of age-old instincts and processes into distinct adaptive actions that offer the greatest chance for survival. For all species, adaptation may involve both conscious and unconscious processes. Examples of conscious, or instinctive process adaptations may be things like herd migration as a species follows changes in its food supply. In the case of *Homo sapiens*, an example of a conscious process adaptation may be the adoption a global initiative to stem climate change. Unconscious adaptations on the other hand involve genetic mutation and evolution. These occurrences are continual and experimental in nature.

What will surely not lead to a specie's long-term survival is for it to stubbornly cling to ancient lore as depicted in symbolic literature. Such dogmatic idealism can only lead a species down a road paved with ignorance and powered by a destructively myopic vision of the future. Dismissal of advances in scientific knowledge and understanding as anti-God and a threat to religion completely misses the wonder and magnificence that our expanding knowledge reveals about creation and the creator.

The plan, if any, that the creator has for *Homo sapiens* is for continued evolution and ultimate extinction. How quickly we come to the end of our evolutionary branch depends, in large part, on how we take care of each other and this tiny, yet wonderful planet. Whether or not our branch bursts forth a new bud or ends in a withered and dry twig is to some degree up to us. We share a common biology and chemistry with countless other living creatures past, present and future. Each one of these living creatures is traveling with us along their own evolutionary branch toward the unknown.

In contrast to a belief in the written and presumed inspired word of God contained in holy books, I believe that the words from the creator are etched deeply, uniformly, universally and in-errantly in the language of creation itself. Man does not need bronze age symbolic literature to develop an understanding of our place in creation or our relationship with the creator. Unlike the holy texts of the world's religions, creation contains no myths. Creation is not flawed by falsehoods or ideological interpretation that can warp human minds into fanatical beliefs and actions. Creation is universal, both in its language and its message. Creation provides a unifying starting point from which human beings can engage in a meaningful discussion about the existence of a creator. Life, and the universe in which we find ourselves offer a common focal point for all people to look upon with wonderment and ask the most basic all human questions. How did all this come to be? Why we are here? Where are we going?

I believe in a creator because I understand a small bit about how life functions and evolves. I cannot accept that the processes necessary to produce energy and reproduce simply happened even over billions of years purely by random events. I believe that life is ever changing. By its very nature, life is committed to a path of perpetual experimentation and change at the molecular level. Is a God directing these changes? Perhaps, but there is no way of knowing if this is so.

Ultimately, there are real explanations for many things we don't understand. Truthful explanations of the unknown don't need to invoke mystery, magic, divine intervention or blind acceptance because *"that's just the way it is"*. Many questions about the mysteries of life and creation have been solved through painstaking research and thoughtful investigation. In many cases, answers to these questions provide us with an astounding and surprising glimpse into the elegant and powerful creative forces upon which life and our universe depend. Still other questions wait for answers yet to be discovered while other questions have yet to have been asked. Science can never hope to provide all answers to all questions for science itself is only capable of posing questions that exist within the realm of human comprehension. A deep reverence and profound wonder for the unknown, in concert with the ceaseless, thoughtful and careful exploration of creation's endless mysteries should be the characteristics that most uniquely distinguish *Homo sapiens* from all other creatures on earth.

Additional Suggested Reading

The following is brief list of additional suggested reading. I found these books and articles particularly thought-provoking, helpful and constructive in writing *Splitting Creation*. Each provides a unique perspective on creation, evolution, religion, faith and humanity.

Armstrong, K. 2007. *The Great Transformation: The Beginning of Our Religious Traditions.* New York: Anchor Books

Armstrong, K. 2009. *The Case for God.* New York: Alfred A. Knopf

Berger, L.R. 2000. *In the Footsteps of Eve: The Mystery of Human Origins.* Washington DC: National Geographic

Carroll, S.B. 2005. *The New Science of Evo Devo: Endless Forms Most Beautiful.* New York: W.W. Norton

Collins, F.S. 2006. *The Language of God.* New York: The Free Press

Darwin, C. 1997. *The Voyage of the Beagle.* Hertfordshire: Wordsworth Editions Ltd

Dawkins, R. 2006. *The God Delusion.* Boston: Houghton Mifflin

Dennett, D.C. 2006. *Breaking the Spell.* New York: The Penguin Group

Gribbin, J. and Gribbin, M. 2003. *How Far is Up? Measuring the Size of the Universe.* Cambridge England: Icon Books

Gugliotta, G. 2008. The Great Human Migration: Why Humans Left Their African homeland 80,000 Years Ago to Colonize the World. *Smithsonian,* 39: 56 - 65.

Hancock, G. 1995. *Finger-Prints of the Gods: The Evidence of Earth's Lost Civilization.* New York: Three Rivers Press

Harris, S. 2004. *The End of Faith.* New York: W.W. Norton

Harris, S. 2006. *Letter to a Christian Nation.* New York: Alfred A. Knopf

Hauser, M.D. 2006. *Moral Minds: How Nature Designed our Universal Sense of Right and Wrong.* New York: HarperCollins

Hayden, T. 2009. What Darwin Didn't Know. *Smithsonian,* 39: 40 - 48.

Kurtz, P. 2007. *Science and Ethics: Can Science Help us Make Wise Moral Judgments?* Amherst, NY: Prometheus Books

Prothero, D.R. 2007. *Evolution: What the Fossils Say and Why it Matters.* New York: Columbia University Press

Quammen, D. 2006. *The Reluctant Mr. Darwin.* New York: W.W. Norton

Quammen, D. 2009. Darwin's First Clues. *National Geographic,* 215: 34 - 55.

Ridley, M. 2009. Modern Darwins. *National Geographic,* 215: 56 - 73.

Roughgarden, J. 2006. *Evolution and Christian Faith.* Washington DC: Island Press

Schroeder, G.L. 1997. *The Science of God: The Convergence of Scientific and Biblical Wisdom.* New York: The Free Press

Stenger, V.J. 2007. *God the Failed Hypothesis: How Science Shows that God does not Exist.* Amherst, NY: Prometheus Books

Weiss, R. 2005. The Stem Cell Divide. *National Geographic.* 208: 2 - 27

Wilson, D.S. 2007. *Evolution for Everyone.* New York: Bantam Dell

Wilson, E.O. 2006. *The Creation: An Appeal to Save Life on Earth.* New York: W.W. Norton

Wolpert, L. 2006. *Six Impossible Things Before Breakfast.* New York: W.W. Norton

About the Author

Erastus Buckrod holds a doctorate in pharmacology and conducted exploratory and clinical research during a 35 year career in the pharmaceutical industry. Erastus also taught pharmacology and physiology to medical and nursing students and led research and graduate programs as a tenured faculty member at a university in the midwestern United States. He retired from professional life in 2008 after holding senior executive positions at a number of pharmaceutical and biotechnology companies. *Splitting Creation* is Erastus Buckrod's first novel.

www.ingramcontent.com/pod-product-compliance
Lightning Source LLC
Chambersburg PA
CBHW032356040426
42451CB00006B/35